Suzuki TS50 X
Service and Repair Manual

by Richard Stone

Models covered
TS50 XKE. 49cc. March 1984 to May 1986
TS50 XKG. 49cc. May 1986 to April 1987
TS50 XKH. 49cc. April 1987 to October 1988
TS50 XKJ. 49cc. October 1988 to March 1993
TS50 XKM. 49cc. August 1991 to January 1994
TS50 XKR. 49cc. February 1994 on

(1599 - 3X5)

© Haynes Group Limited 1999

A book in the **Haynes Service and Repair Manual Series**

ISBN 978 1 85960 526 4

British Library Cataloguing in Publication Data
A catalogue record of this book is available from the British Library

Haynes Group Limited
Haynes North America, Inc

www.haynes.com

Restoring and Preserving our Motoring Heritage

Few people can have had the luck to realise their dreams to quite the same extent and in such a remarkable fashion as John Haynes, Founder and Chairman of the Haynes Publishing Group.

Since 1965 his unique approach to workshop manual publishing has proved so successful that millions of Haynes Manuals are now sold every year throughout the world, covering literally thousands of different makes and models of cars, vans and motorcycles.

A continuing passion for cars and motoring led to the founding in 1985 of a Charitable Trust dedicated to the restoration and preservation of our motoring heritage. To inaugurate the new Museum, John Haynes donated virtually his entire private collection of 52 cars.

Now with an unrivalled international collection of over 210 veteran, vintage and classic cars and motorcycles, the Haynes Motor Museum in Somerset is well on the way to becoming one of the most interesting Motor Museums in the world.

A 70 seat video cinema, a cafe and an extensive motoring bookshop, together with a specially constructed one kilometre motor circuit, make a visit to the Haynes Motor Museum a truly unforgettable experience.

Every vehicle in the museum is preserved in as near as possible mint condition and each car is run every six months on the motor circuit.

Enjoy the picnic area set amongst the rolling Somerset hills. Peer through the William Morris workshop windows at cars being restored, and browse through the extensive displays of fascinating motoring memorabilia.

From the 1903 Oldsmobile through such classics as an MG Midget to the mighty 'E' type Jaguar, Lamborghini, Ferrari Berlinetta Boxer, and Graham Hill's Lola Cosworth, there is something for everyone, young and old alike, at this Somerset Museum.

Haynes Motor Museum

Situated mid-way between London and Penzance, the Haynes Motor Museum is located just off the A303 at Sparkford, Somerset (home of the Haynes Manual) and is open to the public 7 days a week all year round, except Christmas Day and Boxing Day.

Telephone 01963 440804.

Acknowledgements

Our thanks are due to Fran Ridewood & Co of Wells, Somerset, who provided the TS50 XKJ featured in the photographs throughout this manual. The TS50 XKR photograph on the rear cover was supplied by Redcat Marketing.

The Avon Rubber Company supplied information on tyre care and fitting, and NGK Spark Plugs (UK) Ltd. provided information on plug maintenance and electrode conditions.

Contents

The TS 50 XKE model

The TS50 XKJ model

About this manual

The purpose of this manual is to present the owner with a concise and graphic guide which will enable him to tackle any operation from basic routine maintenance to a major overhaul. It has been assumed that any work would be undertaken without the luxury of a well-equipped workshop and a range of manufacturer's service tools.

To this end, the machine featured in the manual was stripped and rebuilt in our own workshop, by a team comprising a mechanic, a photographer and the author. The resulting photographic sequence depicts events as they took place, the hands shown being those of the author and the mechanic.

The use of specialised, and expensive, service tools was avoided unless their use was considered to be essential due to risk of breakage or injury. There is usually some way of improvising a method of removing a stubborn component, providing that a suitable degree of care is exercised.

The author learnt his motorcycle mechanics over a number of years, faced with the same difficulties and using similar facilities to those encountered by most owners. It is hoped that this practical experience can be passed on through the pages of this manual.

Where possible, a well-used example of the machine is chosen for the workshop project, as this highlights any areas which might be particularly prone to giving rise to problems. In this way, any such difficulties are encountered and resolved before the text is written, and the techniques used to deal with them can be incorporated in the relevant section. Armed with a working knowledge of the machine, the author undertakes a considerable amount of research in order that the maximum amount of data can be included in the manual.

A comprehensive section, preceding the main part of the manual, describes procedures for carrying out the routine maintenance of the machine at intervals of time and mileage. This section is included particularly for those owners who wish to ensure the efficient day-to-day running of their motorcycle, but who choose not to undertake overhaul or renovation work.

Each Chapter is divided into numbered sections. Within these sections are numbered paragraphs. Cross reference throughout the manual is quite straightforward and logical. When reference is made 'See Section 6.10' it means Section 6, paragraph 10 in the same Chapter. If another Chapter were intended, the reference would read, for example, 'See Chapter 2, Section 6.10'. All the photographs are captioned with a section/paragraph number to which they refer and are relevant to the Chapter text adjacent.

Figures (usually line illustrations) appear in a logical but numerical order, within a given Chapter. Fig. 1.1 therefore refers to the first figure in Chapter 1.

Left-hand and right-hand descriptions of the machines and their components refer to the left and right of a given machine when the rider is seated normally.

Motorcycle manufacturers continually make changes to specifications and recommendations, and these, when notified, are incorporated into our manuals at the earliest opportunity.

Introduction to the Suzuki TS50 X

Suzuki began producing 50cc trail bikes back in April 1979. The first TS50 followed the popular trail bike styling trends, being fitted with knobbly tyres, high-mounted mudguards, braced handlebars and a high level exhaust. Its successor, the TS50 ER, was introduced in 1980 and became one of the most popular machines with a restricted engine output to comply with legal requirements on the market at that time. During 1984 the TS50 ER was superseded by the TS50 X. Based on the styling of Suzuki's successful RM motocross range, the TS50 X is a complete redesign from its predecessor, with 12 volt electrics, a maintenance-free CDI ignition system and a pear shaped tank to conform to today's styling demands. The rear suspension is Suzuki's own Full Floater type similar in concept to that which many other manufacturers now use as a basis for their own designs. Plastics play a large part in the way of cosmetics, being used for the headlamp cowling, side-panels and mudguards. Full trail bike size wheels are fitted which, when added to the long travel suspension, results in a machine which requires a second glance to confirm that it is indeed a moped.

In common with all other motorcycle manufacturers, Suzuki pursue a policy of constant development during a model's life, ensuring that any modifications or alterations in specification reach the purchaser as soon as possible. Such a policy can, however, lead to some confusion, especially when purchasing replacement parts. It is essential that the machine is identified by its full and correct model designation, which can only be found by using the frame number. To assist the owner in identifying his machine correctly, the following information has been compiled.

Model	Year	Frame number
TS50 XKE	Mar '84 to May '86	SA11C-100001 on
TS50 XKG	May '86 to Apr '87	SA11C-110439 on
TS50 XKH	Apr '87 to Oct '88	SA11C-115476 on
TS50 XKJ	Oct '88 to Mar '93	SA11C-121946 on
TS50 XKM	Aug '91 to Jan '94	SA11C-132261 on
TS50 XKR	Feb '94 on	SA11C-143371 on

Model dimensions and weights

	TS50 XKE	TS50 XKG, H, J, M, R
Overall length	2060 mm (81.2 in)	2140 mm (84.3 in)
Overall width	800 mm (31.5 in)	745 mm (29.3 in)
Overall height	1100 mm (43.3 in)	1155 mm (45.5 in)
Wheelbase	1250 mm (49.3 in)	1340 mm (52.8 in)
Dry weight	83 kg (183 lb)	88 kg (194 lb)
Ground clearance	175 mm (6.9 in)	300 mm (11.8 in)

Ordering spare parts

Before attempting any overhaul or maintenance work it is important to ensure that any parts likely to be required are at hand. Many of the more common parts such as gaskets and seals will be available off the shelf but it will often prove necessary to order any specialised parts in advance. It is well worth running through the operation to be undertaken, referring to the appropriate Chapter and Section of this book, so that a note can be made of the items most likely to be required. In some cases it will, of course, be necessary to dismantle the assembly in question so that the various components can be examined and measured for wear, In these instances, it must be remembered that the machine may have to be left dismantled while the replacement parts are obtained.

When ordering parts for any Suzuki, it is advisable to deal with an authorized Suzuki dealer as parts cannot be obtained direct from the manufacturer. Retain any worn or broken parts until the replacements are at hand; they may be needed as a pattern to help identify the correct replacement when design changes have been made during a production run.

Beware of pattern parts which are usually offered at a lower price to the Suzuki originals. There is no guarantee of their standard and some more important parts have been known to fail suddenly, causing extensive damage.

Some of the more expendable items such as oils, spark plugs, tyres and bulbs can be obtained from accessory shops and motor factors who have convenient opening hours and can usually be found closer to home. It is also possible to obtain parts on a mail order basis from a number of specialists who advertise regularly in the motorcycle press.

Most machines are subject to continuous detail modifications during their production run. These developments will be unknown to the general public so it is essential to quote the engine and frame numbers in full when ordering parts. The engine number is stamped on a raised bar on the crankcase left-hand half and the frame number is stamped on the steering head.

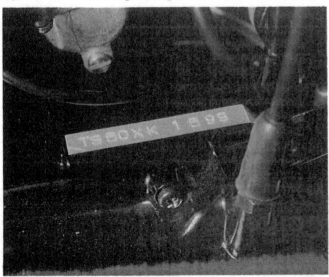

Engine number location

Frame number location

Safety first!

Professional motor mechanics are trained in safe working procedures. However enthusiastic you may be about getting on with the job in hand, do take the time to ensure that your safety is not put at risk. A moment's lack of attention can result in an accident, as can failure to observe certain elementary precautions.

There will always be new ways of having accidents, and the following points do not pretend to be a comprehensive list of all dangers; they are intended rather to make you aware of the risks and to encourage a safety-conscious approach to all work you carry out on your vehicle.

Essential DOs and DON'Ts

DON'T start the engine without first ascertaining that the transmission is in neutral.

DON'T suddenly remove the filler cap from a hot cooling system – cover it with a cloth and release the pressure gradually first, or you may get scalded by escaping coolant.

DON'T attempt to drain oil until you are sure it has cooled sufficiently to avoid scalding you.

DON'T grasp any part of the engine, exhaust or silencer without first ascertaining that it is sufficiently cool to avoid burning you.

DON'T allow brake fluid or antifreeze to contact the machine's paintwork or plastic components.

DON'T syphon toxic liquids such as fuel, brake fluid or antifreeze by mouth, or allow them to remain on your skin.

DON'T inhale dust – it may be injurious to health (see *Asbestos* heading).

DON'T allow any spilt oil or grease to remain on the floor – wipe it up straight away, before someone slips on it.

DON'T use ill-fitting spanners or other tools which may slip and cause injury.

DON'T attempt to lift a heavy component which may be beyond your capability – get assistance.

DON'T rush to finish a job, or take unverified short cuts.

DON'T allow children or animals in or around an unattended vehicle.

DON'T inflate a tyre to a pressure above the recommended maximum. Apart from overstressing the carcase and wheel rim, in extreme cases the tyre may blow off forcibly.

DO ensure that the machine is supported securely at all times. This is especially important when the machine is blocked up to aid wheel or fork removal.

DO take care when attempting to slacken a stubborn nut or bolt. It is generally better to pull on a spanner, rather than push, so that if slippage occurs you fall away from the machine rather than on to it.

DO wear eye protection when using power tools such as drill, sander, bench grinder etc.

DO use a barrier cream on your hands prior to undertaking dirty jobs – it will protect your skin from infection as well as making the dirt easier to remove afterwards; but make sure your hands aren't left slippery. Note that long-term contact with used engine oil can be a health hazard.

DO keep loose clothing (cuffs, tie etc) and long hair well out of the way of moving mechanical parts.

DO remove rings, wristwatch etc, before working on the vehicle – especially the electrical system.

DO keep your work area tidy – it is only too easy to fall over articles left lying around.

DO exercise caution when compressing springs for removal or installation. Ensure that the tension is applied and released in a controlled manner, using suitable tools which preclude the possibility of the spring escaping violently.

DO ensure that any lifting tackle used has a safe working load rating adequate for the job.

DO get someone to check periodically that all is well, when working alone on the vehicle.

DO carry out work in a logical sequence and check that everything is correctly assembled and tightened afterwards.

DO remember that your vehicle's safety affects that of yourself and others. If in doubt on any point, get specialist advice.

IF, in spite of following these precautions, you are unfortunate enough to injure yourself, seek medical attention as soon as possible.

Asbestos

Certain friction, insulating, sealing, and other products – such as brake linings, clutch linings, gaskets, etc – contain asbestos. *Extreme care must be taken to avoid inhalation of dust from such products since it is hazardous to health.* If in doubt, assume that they *do* contain asbestos.

Fire

Remember at all times that petrol (gasoline) is highly flammable. Never smoke, or have any kind of naked flame around, when working on the vehicle. But the risk does not end there – a spark caused by an electrical short-circuit, by two metal surfaces contacting each other, by careless use of tools, or even by static electricity built up in your body under certain conditions, can ignite petrol vapour, which in a confined space is highly explosive.

Always disconnect the battery earth (ground) terminal before working on any part of the fuel or electrical system, and never risk spilling fuel on to a hot engine or exhaust.

It is recommended that a fire extinguisher of a type suitable for fuel and electrical fires is kept handy in the garage or workplace at all times. Never try to extinguish a fuel or electrical fire with water.

Note: *Any reference to a 'torch' appearing in this manual should always be taken to mean a hand-held battery-operated electric lamp or flashlight. It does **not** mean a welding/gas torch or blowlamp.*

Fumes

Certain fumes are highly toxic and can quickly cause unconsciousness and even death if inhaled to any extent. Petrol (gasoline) vapour comes into this category, as do the vapours from certain solvents such as trichloroethylene. Any draining or pouring of such volatile fluids should be done in a well ventilated area.

When using cleaning fluids and solvents, read the instructions carefully. Never use materials from unmarked containers – they may give off poisonous vapours.

Never run the engine of a motor vehicle in an enclosed space such as a garage. Exhaust fumes contain carbon monoxide which is extremely poisonous; if you need to run the engine, always do so in the open air or at least have the rear of the vehicle outside the workplace.

The battery

Never cause a spark, or allow a naked light, near the vehicle's battery. It will normally be giving off a certain amount of hydrogen gas, which is highly explosive.

Always disconnect the battery earth (ground) terminal before working on the fuel or electrical systems.

If possible, loosen the filler plugs or cover when charging the battery from an external source. Do not charge at an excessive rate or the battery may burst.

Take care when topping up and when carrying the battery. The acid electrolyte, even when diluted, is very corrosive and should not be allowed to contact the eyes or skin.

If you ever need to prepare electrolyte yourself, always add the acid slowly to the water, and never the other way round. Protect against splashes by wearing rubber gloves and goggles.

Mains electricity and electrical equipment

When using an electric power tool, inspection light etc, always ensure that the appliance is correctly connected to its plug and that, where necessary, it is properly earthed (grounded). Do not use such appliances in damp conditions and, again, beware of creating a spark or applying excessive heat in the vicinity of fuel or fuel vapour. Also ensure that the appliances meet the relevant national safety standards.

Ignition HT voltage

A severe electric shock can result from touching certain parts of the ignition system, such as the HT leads, when the engine is running or being cranked, particularly if components are damp or the insulation is defective. Where an electronic ignition system is fitted, the HT voltage is much higher and could prove fatal.

Tools and working facilities

The first priority when undertaking maintenance or repair work of any sort on a motorcycle is to have a clean, dry, well-lit working area. Work carried out in peace and quiet in the well-ordered atmosphere of a good workshop will give more satisfaction and much better results than can usually be achieved in poor working conditions. A good workshop must have a clean flat workbench or a solidly constructed table of convenient working height. The workbench or table should be equipped with a vice which has a jaw opening of at least 4 in (100 mm). A set of jaw covers should be made from soft metal such as aluminium alloy or copper, or from wood. These covers will minimise the marking or damaging of soft or delicate components which may be clamped in the vice. Some clean, dry, storage space will be required for tools, lubricants and dismantled components. It will be necessary during a major overhaul to lay out engine/gearbox components for examination and to keep them where they will remain undisturbed for as long as is necessary. To this end it is recommended that a supply of metal or plastic containers of suitable size is collected. A supply of clean, lint-free, rags for cleaning purposes and some newspapers, other rags, or paper towels for mopping up spillages should also be kept. If working on a hard concrete floor note that both the floor and one's knees can be protected from oil spillages and wear by cutting open a large cardboard box and spreading it flat on the floor under the machine or workbench. This also helps to provide some warmth in winter and to prevent the loss of nuts, washers, and other tiny components which have a tendency to disappear when dropped on anything other than a perfectly clean, flat surface.

Unfortunately, such working conditions are not always available to the home mechanic. When working in poor conditions it is essential to take extra time and care to ensure that the components being worked on are kept scrupulously clean and to ensure that no components or tools are lost or damaged.

A selection of good tools is a fundamental requirement for anyone contemplating the maintenance and repair of a motor vehicle. For the owner who does not possess any, their purchase will prove a considerable expense, offsetting some of the savings made by doing-it-yourself. However, provided that the tools purchase meet the relevant national safety standards and are of good quality, they will last for many years and prove an extremely worthwhile investment.

To help the average owner to decide which tools are needed to carry out the various tasks detailed in this manual, we have compiled three lists of tools under the following headings: *Maintenance and minor repair*, *Repair and overhaul*, and *Specialized*. The newcomer to practical mechanics should start off with the simpler jobs around the vehicle. Then, as his confidence and experience grow, he can undertake the difficult tasks, buying extra tools as and when they are needed. In this way, a *Maintenance and minor repair* tool kit can be built-up into a *Repair and overhaul* tool kit over a considerable period of time without any major cash outlays. The experienced home mechanic will have a tool kit good enough for most repair and overhaul procedures and will add tools from the specialized category when he feels the expense is justified by the amount of use these tools will be put to.

It is obviously not possible to cover the subject of tools fully here. For those who wish to learn more about tools and their use there is a book entitled Motorcycle Workshop Practice Techbook (Book no 3470) available from the publishers of this manual.

As a general rule, it is better to buy the more expensive, good quality tools. Given reasonable use, such tools will last for a very long time, whereas the cheaper, poor quality, items will wear out faster and need to be renewed more often, thus nullifying the original saving. There is also the risk of a poor quality tool breaking while in use, causing personal injury or expensive damage to the component being worked on. For practically all tools, a tool factor is the best source since he will have a very comprehensive range compared with the average garage or accessory shop. Having said that, accessory shops often offer excellent quality tools at discount prices, so it pays to shop around. There are plenty of tools around at reasonable prices, but always aim to purchase items which meet the relevant national safety standards. If in doubt, seek the advice of the shop proprietor or manager before making a purchase.

The basis of any toolkit is a set of spanners. While open-ended spanners with their slim jaws, are useful for working on awkwardly-positioned nuts, ring spanners have advantages in that they grip the nut far more positively. There is less risk of the spanner slipping off the nut and damaging it, for this reason alone ring spanners are to be preferred. Ideally, the home mechanic should acquire a set of each, but if expense rules this out a set of combination spanners (open-ended at one end and with a ring of the same size at the other) will provide a good compromise. Another item which is so useful it should be considered an essential requirement for any home mechanic is a set of socket spanners. These are available in a variety of drive sizes. It is recommended that the ½-inch drive type is purchased to begin with as although bulkier and more expensive than the ⅜-inch type, the larger size is far more common and will accept a greater variety of torque wrenches, extension pieces and socket sizes. The socket set should comprise sockets of sizes between 8 and 24 mm, a reversible ratchet drive, an extension bar of about 10 inches in length, a spark plug socket with a rubber insert, and a universal joint. Other attachments can be added to the set at a later date.

Maintenance and minor repair tool kit

Set of spanners 8 – 24 mm
Set of sockets and attachments
Spark plug spanner with rubber insert – 10, 12, or 14 mm as appropriate
Adjustable spanner
C-spanner/pin spanner
Torque wrench (same size drive as sockets)
Set of screwdrivers (flat blade)
Set of screwdrivers (cross-head)
Set of Allen keys 4 – 10 mm
Impact screwdriver and bits
Ball pein hammer – 2 lb
Hacksaw (junior)
Self-locking pliers – Mole grips or vice grips
Pliers – combination
Pliers – needle nose
Wire brush (small)
Soft-bristled brush
Tyre pump
Tyre pressure gauge
Tyre tread depth gauge

Oil can
Fine emery cloth
Funnel (medium size)
Drip tray
Grease gun
Set of feeler gauges
Strobe timing light
Continuity tester (dry battery and bulb)
Soldering iron and solder
Wire stripper or craft knife
PVC insulating tape
Assortment of split pins, nuts, bolts, and washers

Repair and overhaul toolkit

The tools in this list are virtually essential for anyone undertaking major repairs to a motorcycle and are additional to the tools listed above.

Plastic or rubber soft-faced mallet
Pliers – electrician's side cutters
Circlip pliers – internal (straight or right-angled tips are available)
Circlip pliers – external
Cold chisel
Centre punch
Pin punch
Scriber
Scraper (made from soft metal such as aluminium or copper)
Soft metal drift
Steel rule/straightedge
Assortment of files
Electric drill and bits
Wire brush (large)
Soft wire brush (similar to those used for cleaning suede shoes)
Sheet of plate glass
Hacksaw (large)
Stud extractor set (E-Z out)

Specialized tools

This is not a list of the tools made by the machine's manufacturer to carry out a specific task on a limited range of models. Occasional references are made to such tools in the text of this manual and, in general, an alternative method of carrying out the task without the manufacturer's tool is given where possible. The tools mentioned in this list are those which are not used regularly and are expensive to buy in view of their infrequent use. Where this is the case it may be possible to hire or borrow the tools against a deposit from a local dealer or tool hire shop. An alternative is for a group of friends or a motorcycle club to join in the purchase.

Piston ring compressor
Universal bearing puller
Cylinder bore honing attachment (for electric drill)
Micrometer set
Vernier calipers
Dial gauge set
Cylinder compression gauge
Multimeter

Care and maintenance of tools

Whatever the quality of the tools purchased, they will last much longer if cared for. This means in practice ensuring that a tool is used for its intended purpose; for example screwdrivers should not be used as a substitute for a centre punch, or as chisels. Always remove dirt or grease and any metal particles but remember that a light film of oil will prevent rusting if the tools are infrequently used. The common tools can be kept together in a large box or tray but the more delicate, and more expensive, items should be stored separately where they cannot be damaged. When a tool is damaged or worn out, be sure to renew it immediately. It is false economy to continue to use a worn spanner or screwdriver which may slip and cause expensive damage to the component being worked on.

Fastening systems

Fasteners, basically, are nuts, bolts and screws used to hold two or more parts together. There are a few things to keep in mind when working with fasteners. Almost all of them use a locking device of some type; either a lock washer, locknut, locking tab or thread adhesive. All threaded fasteners should be clean, straight, have undamaged threads and undamaged corners on the hexagon head where the spanner fits. Develop the habit of replacing all damaged nuts and bolts with new ones.

Rusted nuts and bolts should be treated with a rust penetrating fluid to ease removal and prevent breakage. After applying the rust penetrant, let it 'work' for a few minutes before trying to loosen the nut or bolt. Badly rusted fasteners may have to be chiseled off or removed with a special nut breaker, available at tool shops.

Flat washers and lock washers, when removed from an assembly should always be replaced exactly as removed. Replace any damaged washers with new ones. Always use a flat washer between a lock washer and any soft metal surface (such as aluminium), thin sheet metal or plastic. Special locknuts can only be used once or twice before they lose their locking ability and must be renewed.

If a bolt or stud breaks off in an assembly, it can be drilled out and removed with a special tool called an E-Z out. Most dealer service departments and motorcycle repair shops can perform this task, as well as others (such as the repair of threaded holes that have been stripped out).

Spanner size comparison

Jaw gap (in)	Spanner size	Jaw gap (in)	Spanner size
0.250	1/4 in AF	0.945	24 mm
0.276	7 mm	1.000	1 in AF
0.313	5/16 in AF	1.010	9/16 in Whitworth; 5/8 in BSF
0.315	8 mm	1.024	26 mm
0.344	11/32 in AF; 1/8 in Whitworth	1.063	11/16 in AF; 27 mm
0.354	9 mm	1.100	5/16 in Whitworth; 11/16 in BSF
0.375	3/8 in AF	1.125	11/8 in AF
0.394	10 mm	1.181	30 mm
0.433	11 mm	1.200	11/16 in Whitworth; 3/4 in BSF
0.438	7/16 in AF	1.250	11/4 in AF
0.445	3/16 in Whitworth; 1/4	1.260	32 mm
0.472	12 mm	1.300	3/4 in Whitworth; 7/8 in BSF
0.500	1/2 in AF	1.313	15/16 in AF
0.512	13 mm	1.390	13/16 in Whitworth; 15/16 in BSF
0.525	1/4 in Whitworth; 5/16 in BSF	1.417	36 mm
0.551	14 mm	1.438	17/16 in AF
0.563	9/16 in AF	1.480	7/8 in Whitworth; 1 in BSF
0.591	15 mm	1.500	11/2 in AF
0.600	5/16 in Whitworth; 3/8 in BSF	1.575	40 mm; 15/16 in Whitworth
0.625	5/8 in AF	1.614	41 mm
0.630	16 mm	1.625	15/8 in AF
0.669	17 mm	1.670	1 in Whitworth; 11/8 in BSF
0.686	11/16 in AF	1.688	111/16 in AF
0.709	18 mm	1.811	46 mm
0.710	3/8 in Whitworth; 7/16 in BSF	1.813	113/16 in AF
0.748	19 mm	1.860	11/8 in Whitworth; 11/4 in BSF
0.750	3/4 in AF	1.875	17/8 in AF
0.813	13/16 in AF	1.969	50 mm
0.820	7/16 in Whitworth; 1/2 in BSF	2.000	2 in AF
0.866	22 mm	2.050	11/4 in Whitworth; 13/8 in BSF
0.875	7/8 in AF	2.165	55 mm
0.920	1/2 in Whitworth; 9/16 in BSF	2.362	60 mm
0.938	15/16 in AF		

Standard torque settings

Specific torque settings will be found at the end of the specifications section of each chapter. Where no figure is given, it should be secured according to the table below.

Fastener type (thread diameter)	kgf m	lbf ft
5mm bolt or nut	0.45 – 0.6	3.5 – 4.5
6 mm bolt or nut	0.8 – 1.2	6 – 9
8 mm bolt or nut	1.8 – 2.5	13 – 18
10 mm bolt or nut	3.0 – 4.0	22 – 29
12 mm bolt or nut	5.0 – 6.0	36 – 43
5 mm screw	0.35 – 0.5	2.5 – 3.6
6 mm screw	0.7 – 1.1	5 – 8
6 mm flange bolt	1.0 – 1.4	7 – 10
8 mm flange bolt	2.4 – 3.0	17 – 22
10 mm flange bolt	3.5 – 4.5	25 – 33

Fault diagnosis

Contents

1 Introduction

This Section provides an easy reference-guide to the more common faults that are likely to afflict your machine. Obviously, the opportunities are almost limitless for faults to occur as a result of obscure failures, and to try and cover all eventualities would require a book. Indeed, a number have been written on the subject.

Successful fault diagnosis is not a mysterious 'black art' but the application of a bit of knowledge combined with a systematic and logical approach to the problem. Approach any fault diagnosis by first accurately identifying the symptom and then checking through the list of possible causes, starting with the simplest or most obvious and progressing in stages to the most complex. Take nothing for granted, but above all apply liberal quantities of common sense.

The main symptom of a fault is given in the text as a major heading below which are listed, as Section headings, the various systems or areas which may contain the fault. Details of each possible cause for a fault and the remedial action to be taken are given, in brief, in the paragraphs below each Section heading. Further information should be sought in the relevant Chapter.

Engine does not start when turned over

2 No fuel flow to carburettor

● Fuel tank empty or level too low. Check that the tap is turned to 'On' or 'Reserve' position as required. If in doubt, prise off the fuel feed pipe at the carburettor end and check that fuel runs from the pipe when the tap is turned on.
● Tank filler cap vent pipe obstructed or trapped. This can prevent fuel

from flowing into the carburettor float bowl because air cannot enter the fuel tank to replace it. The problem is more likely to appear when the machine is being ridden. Check by listening close to the filler cap and releasing it. A hissing noise indicates that a blockage is present. Remove the cap and clear the vent pipe.
● Fuel tap or filter blocked. Blockage may be due to accumulation of rust or paint flakes from the tank's inner surface or of foreign matter from contaminated fuel. Remove the tap and clean it and the filter. Look also for water droplets in the fuel.
● Fuel line blocked. Blockage of the fuel line is more likely to result from a kink in the line rather than the accumulation of debris.

3 Fuel not reaching cylinder

● Float chamber not filling. Caused by float needle or floats sticking in up position. This may occur after the machine has been left standing for an extended length of time allowing the fuel to evaporate. When this occurs a gummy residue is often left which hardens to a varnish-like substance. This condition may be worsened by corrosion and crystalline deposits produced prior to the total evaporation of contaminated fuel. Sticking of the float needle may also be caused by wear. In any case removal of the float chamber will be necessary for inspection and cleaning.
● Blockage in starting circuit, slow running circuit or jets. Blockage of these items may be attributable to debris from the fuel tank by-passing the filter system or to gumming up as described in paragraph 1. Water droplets in the fuel will also block jets and passages. The carburettor should be dismantled for cleaning.
● Fuel level too low. The fuel level in the float chamber is controlled by float height. The fuel level may increase with wear or damage but will never reduce, thus a low fuel level is an inherent rather than developing condition. Check the float height and adjust if necessary.

4 Engine flooding

● Float valve needle worn or stuck open. A piece of rust or other debris can prevent correct seating of the needle against the valve seat thereby permitting an uncontrolled flow of fuel. Similarly, a worn needle or needle seat will prevent valve closure. Dismantle the carburettor float bowl for cleaning and, if necessary, renewal of the worn components.
● Fuel level too high. The fuel level is controlled by the float height which may increase due to wear of the float needle, pivot pin or operating tang. Check the float height, and make any necessary adjustments. A leaking float will cause an increase in fuel level, and thus should be renewed.
● Cold starting mechanism. Check the choke (starter mechanism) for correct operation. If the mechanism jams in the 'On' position subsequent starting of a hot engine will be difficult.
● Blocked air filter. A badly restricted air filter will cause flooding. Check the filter and clean or renew as required. A collapsed inlet hose will have a similar effect. Check that the air filter inlet has not become blocked by a rag or similar item.

5 No spark at plug

● Ignition switch not on.
● Engine stop switch off.
● Spark plug dirty, oiled or 'fouled'. Because the induction mixture of a two-stroke engine is inclined to be of a rather oily nature it is comparatively easy to foul the plug electrodes, especially where there have been repeated attempts to start the engine. A machine used for short journeys will be more prone to fouling because the engine may never reach full operating temperature, and the deposits will not burn off. On rare occasions a change of plug grade may be required but the advice of a dealer should be sought before making such a change. An electrode deposit in the form of a barely visible filament across the plug electrodes can short circuit the plug and prevent its sparking. On all two-stroke

machines it is a sound precaution to carry a new spare spark plug for substitution in the event of fouling problems.
● Spark plug failure. Clean the spark plug thoroughly and reset the electrode gap. Refer to the spark plug section
        in the Routine maintenance section. If the spark plug shorts internally or has sustained visible damage to the electrodes, core or ceramic insulator it should be renewed. On rare occasions a plug that appears to spark vigorously will fail to do so when refitted to the engine and subjected to the compression pressure in the cylinder.
● Spark plug cap or high tension (HT) lead faulty. Check condition and security. Replace if deterioration is evident. Most spark plug caps have an internal resistor designed to inhibit electrical interference with radio and television sets. On rare occasions the resistor may break down, thus preventing sparking. If this is suspected, fit a new cap as a precaution.
● Spark plug cap loose. Check that the spark plug cap fits securely over the plug and, where fitted, the screwed terminal on the plug end is secure.
● Shorting due to moisture. Certain parts of the ignition system are susceptible to shorting when the machine is ridden or parked in wet weather. Check particularly the area from the spark plug cap back to the ignition coil. A water dispersant spray may be used to dry out water-logged components. Recurrence of the problem can be prevented by using an ignition sealant spray after drying out and cleaning.
● Ignition or stop switch shorted. May be caused by water corrosion or wear. Water dispersant and contact cleaning sprays may be used. If this fails to overcome the problem dismantling and visual inspection of the switches will be required.
● Shorting or open circuit in wiring. Failure in any wire connecting any of the ignition components will cause ignition malfunction. Check also that all connections are clean, dry and tight.
● Ignition coil/CDI unit failure. Check the coil, referring to Chapter 3.
● Generator source coil failure. Refer to Chapter 3.

6 Weak spark at plug

● Feeble sparking at the plug may be caused by any of the faults mentioned in the preceding Section other than those items in the first two paragraphs. Check first the spark plug, this being the most likely culprit.

7 Compression low

● Spark plug loose. This will be self-evident on inspection, and may be accompanied by a hissing noise when the engine is turned over. Remove the plug and check that the threads in the cylinder head are not damaged. Check also that the plug sealing washer is in good condition.
● Cylinder head gasket leaking. This condition is often accompanied by a high pitched squeak from around the cylinder head and oil loss, and may be caused by insufficiently tightened cylinder head fasteners, a warped cylinder head or mechanical failure of the gasket material. Re-torqueing the fasteners to the correct specification may seal the leak in some instances but if damage has occurred this course of action will provide, at best, only a temporary cure.
● Low crankcase compression. This can be caused by worn main bearings and seals and will upset the incoming fuel/air mixture. A good seal in these areas is essential on any two-stroke engine.
● Piston rings sticking or broken. Sticking of the piston rings may be caused by seizure due to lack of lubrication or overheating as a result of poor carburation or incorrect fuel type. Gumming of the rings may result from lack of use, or carbon deposits in the ring grooves. Broken rings result from over-revving, over-heating or general wear. In either case a top-end overhaul will be required.

Engine stalls after starting

8 General causes

● Improper cold start mechanism operation. Check that the operating controls function smoothly. A cold engine may not require application of

an enriched mixture to start initially but may baulk without choke once firing. Likewise a hot engine may start with an enriched mixture but will stop almost immediately if the choke is inadvertently in operation.
● Ignition malfunction. See Section 6. Weak spark at plug
● Carburettor incorrectly adjusted. Maladjustment of the mixture strength or idle speed may cause the engine to stop immediately after starting. See Chapter 2.
● Fuel contamination. Check for filter blockage by debris or water which reduces, but does not completely stop, fuel flow, or blockage of the slow speed circuit in the carburettor by the same agents. If water is present it can often be seen as droplets in the bottom of the float bowl. Clean the fuel tap filter and, where water is in evidence, drain and flush the fuel tank and float bowl.
● Intake air leak. Check for security of the carburettor mounting and hose connections, and for cracks or splits in the hoses. Check also that the carburettor top is secure.
● Air filter blocked or omitted. A blocked filter will cause an over-rich mixture; the omission of a filter will cause an excessively weak mixture. Both conditions will have a detrimental effect on carburation. Clean or renew the filter as necessary.
● Fuel filler cap air vent pipe blocked or trapped. Usually caused by dirt or water. Clean the vent orifice.
● Choked exhaust system. Caused by excessive carbon build-up in the system. Refer to Chapter 2 for further information.
● Excessive carbon build-up in the engine. This can result from failure to decarbonise the engine at the specified interval or through excessive oil consumption. Check pump adjustment.

Poor running at idle and low speed

9   Weak spark at plug or erratic firing

● Battery voltage low. In certain conditions low battery charge, especially when coupled with a badly sulphated battery, may result in misfiring. If the battery is in good general condition it should be recharged; an old battery suffering from sulphated plates should be renewed.
● Spark plug fouled, faulty or incorrectly adjusted. See Section 5.
● Spark plug cap or high tension lead shorting. Check the condition of both these items ensuring that they are in good condition and dry and that the cap is fitted correctly.
● Spark plug type incorrect. Fit plug of correct type and heat range as given in Specifications. In certain conditions a plug of hotter or colder type may be required for normal running.
● Ignition timing incorrect. Check the ignition timing as described in Chapter 3.
● Faulty ignition coil. Partial failure of the coil internal insulation will diminish the performance of the coil. No repair is possible, a new component must be fitted.
● Defective ignition source coil. Refer to Chapter 3 for further details on test procedures.

10   Fuel/air mixture incorrect

● Intake air leak. Check carburettor mountings and air filter hoses for security and signs of splitting. Ensure that carburettor top is tight.
● Mixture strength incorrect. Adjust mixture strength using pilot adjustment screw.
● Pilot jet or slow running circuit blocked. The carburettor should be removed and dismantled for thorough cleaning. Blow through all jets and air passages with compressed air to clear obstructions.
● Air filter clogged or omitted. Clean or fit air filter element as necessary. Check also that the element and air filter cover are correctly seated.
● Cold start mechanism in operation. Check that the choke has not been left on inadvertently and the operation is correct.
● Fuel level too high or too low. Check the float height and adjust if necessary. See Section 3 or 4.
● Fuel tank air vent pipe obstructed or trapped. Obstructions usually caused by dirt or water.

11   Compression low

● See Section 7.

Acceleration poor

12   General causes

● All items as for previous Section.
● Choked air filter. Failure to keep the air filter element clean will allow the build-up of dirt with proportional loss of performance. In extreme cases of neglect acceleration will suffer.
● Choked exhaust system. This can result from failure to clean the system at the prescribed intervals. The increased back pressure will make the machine noticeably sluggish. Refer to Chapter 2 for further information on decarbonisation.
● Excessive carbon build-up in the engine. This can result from failure to decarbonise the engine at the specified interval or through excessive oil consumption. Check pump adjustment.
● Ignition timing incorrect. Check as described in Chapter 3.
● Carburation fault. See Section 10.
● Mechanical resistance. Check that the brakes are not binding. Note that the increased rolling resistance caused by under-inflated tyres may impede acceleration.

Poor running or lack of power at high speeds

13   Weak spark at plug or erratic firing

● All items as for Section 9.
● HT lead insulation failure. Insulation failure of the HT lead and spark plug cap due to old age or damage can cause shorting when the engine is driven hard. This condition may be less noticeable, or not noticeable at all at lower engine speeds.

14   Fuel/air mixture incorrect

● All items as for Section 10, with the exception of items relative exclusively to low speed running.
● Main jet blocked. Debris from contaminated fuel, or from the fuel tank, and water in the fuel can block the main jet. Clean the fuel tap filter, the float bowl area, and if water is present, flush and refill the fuel tank.
● Main jet is the wrong size. The standard carburettor jetting is for sea level atmospheric pressure. For high altitudes, usually above 5000 ft, a smaller main jet will be required.
● Jet needle and needle jet worn. These can be renewed individually but should be renewed as a pair. Renewal of both items requires partial dismantling of the carburettor.
● Air bleed holes blocked. Dismantle carburettor and use compressed air to blow out all air passages.
● Reduced fuel flow. A reduction in the maximum fuel flow from the fuel tank to the carburettor will cause fuel starvation, proportionate to the engine speed. Check for blockages through debris or a kinked fuel line.

15   Compression low

● See Section 7.

Knocking or pinking

16   General causes

● Carbon build-up in combustion chamber. After high mileages have been covered a large accumulation of carbon may occur. This may glow red hot and cause premature ignition of the fuel/air mixture, in advance of normal firing by the spark plug. Cylinder head removal will be required to allow inspection and cleaning.

● Fuel incorrect. A low grade fuel, or one of poor quality may result in compression induced detonation of the fuel resulting in knocking and pinking noises. Old fuel can cause similar problems. A too highly leaded fuel will reduce detonation but will accelerate deposit formation in the combustion chamber and may lead to early pre-ignition as described in item 1.

● Spark plug heat range incorrect. Uncontrolled pre-ignition can result from the use of a spark plug the heat range of which is too hot.

● Weak mixture. Overheating of the engine due to a weak mixture can result in pre-ignition occurring where it would not occur when engine temperature was within normal limits. Maladjustment, blocked jets or passages and air leaks can cause this condition.

Overheating

17 Firing incorrect

● Spark plug fouled, defective or maladjusted. See Section 5.
● Spark plug type incorrect. Refer to the Specifications and ensure that the correct plug type is fitted.
● Incorrect ignition timing. Timing that is far too much advanced or far too much retarded will cause overheating. Check the ignition timing as described in Chapter 3.

18 Fuel/air mixture incorrect

● Slow speed mixture strength incorrect. Adjust pilot screw.
● Main jet wrong size. The carburettor is jetted for sea level atmospheric conditions. For high altitudes, usually above 5000 ft, a smaller main jet will be required.
● Air filter badly fitted or omitted. Check that the filter element is in place and that it and the air filter box cover are sealing correctly. Any leaks will cause a weak mixture.
● Induction air leaks. Check the security of the carburettor mountings and hose connections, and for cracks and splits in the hoses. Check also that the carburettor top is secure.
● Fuel level too low. See Section 3.
● Fuel tank filler cap air vent pipe obstructed or trapped. Clear blockage.

19 Lubrication inadequate

● Oil pump settings incorrect. The oil pump settings are of great importance since the quantities of oil being injected are very small. Any variation in oil delivery will have a significant effect on the engine. Refer to Chapter 2 for further information.
● Oil tank empty or low. This will have disastrous consequences if left unnoticed. Check and replenish tank regularly.
● Transmission oil low or worn out. Check the level regularly and investigate any loss of oil. If the oil level drops with no sign of external leakage it is likely that the crankshaft main bearing oil seals are worn, allowing transmission oil to be drawn into the crankcase during induction.

20 Miscellaneous causes

● Engine fins clogged. A build-up of mud in the cylinder head and cylinder barrel cooling fins will decrease the cooling capabilities of the fins. Clean the fins as required.

Clutch operating problems

21 Clutch slip

● No clutch lever play. Adjust clutch lever free play according to the procedure in Routine maintenance.
● Friction plates worn or warped. Overhaul clutch assembly, replacing plates out of specification.
● Steel plates worn or warped. Overhaul clutch assembly, replacing plates out of specification.
● Clutch spring broken or worn. Old or heat-damaged (from slipping clutch) springs should be replaced with new ones.
● Clutch release not adjusted properly. See Routine maintenance.
● Clutch inner cable snagging. Caused by a frayed cable or kinked outer cable. Replace the cable with a new one. Repair of a frayed cable is not advised.
● Clutch release mechanism defective. Worn or damaged parts in the clutch release mechanism could include the operating arm, worm drive or pushrod. Renew parts as necessary.
● Clutch centre and outer drum worn. Severe indentation by the clutch plate tangs of the channels in the centre and drum will cause snagging of the plates preventing correct engagement. If this damage occurs, renewal of the worn components is required.
● Lubricant incorrect. Use of a transmission lubricant other than that specified may allow the plates to slip.

22 Clutch drag

● Clutch lever play excessive. Adjust cable as described in Routine maintenance.
● Clutch plates warped or damaged. This will cause a drag on the clutch, causing the machine to creep. Overhaul clutch assembly.
● Clutch spring tension uneven. Usually caused by a sagged or broken spring. Check and replace springs.
● Transmission oil deteriorated. Badly contaminated transmission oil and a heavy deposit of oil sludge on the plates will cause plate sticking. The oil recommended for this machine is of the detergent type, therefore it is unlikely that this problem will arise unless regular oil changes are neglected.
● Transmission oil viscosity too high. Drag in the plates will result from the use of an oil with too high a viscosity. In very cold weather clutch drag may occur until the engine has reached operating temperature.
● Clutch centre and outer drum worn. Indentation by the clutch plate tangs of the channels in the centre and drum will prevent easy plate disengagement. If the damage is light the affected areas may be dressed with a fine file. More pronounced damage will necessitate renewal of the components.
● Clutch outer drum seized to shaft. Lack of lubrication, severe wear or damage can cause the drum to seize to the shaft. Overhaul of the clutch, and perhaps the transmission, may be necessary to repair damage.
● Clutch release mechanism defective. Worn or damaged release mechanism parts can stick and fail to provide leverage. Overhaul release assembly.
● Loose clutch hub nut. Causes drum and centre misalignment, putting a drag on the engine. Engagement adjustment continually varies. Overhaul clutch assembly.

Gear selection problems

23 Gear lever does not return

● Weak or broken return spring. Renew the spring.
● Gearchange shaft bent or seized. Distortion of the gearchange shaft often occurs if the machine is dropped heavily on the gear lever. Provided that damage is not severe, straightening of the shaft is permissible.

24 Gear selection difficult or impossible

● Clutch not disengaging fully. See Section 22.
● Gearchange shaft bent. This often occurs if the machine is dropped heavily on the gear lever. Straightening of the shaft is permissible if the damage is not too great.
● Gearchange ratchet or pawls worn or damaged. Wear or breakage of

any of these items may cause difficulty in selecting one or more gears. Overhaul the selector mechanism.
● Gearchange drum stopper cam damaged. Failure, rather than wear of this item may jam the drum thereby preventing gearchanging or causing false selection at high speed.
● Selector forks bent or seized. This can be caused by dropping the machine heavily on the gearchange lever or as a result of lack of lubrication. Though rare, bending of a shaft can result from a missed gearchange or false selection at high speed.
● Selector fork end and pin wear. Pronounced wear of these items and the grooves in the gearchange drum can lead to imprecise selection and, eventually, no selection. Renewal of the worn components will be required.
● Structural failure. Failure of any one component of the selector rod and change mechanism will result in improper or fouled gear selection.

25 Jumping out of gear

● Detent assembly worn or damaged. Wear of the stopper arm and the cam with which it locates and breakage of the stopper arm spring can cause imprecise gear selection resulting in jumping out of gear. Renew the damaged components.
● Gear pinion dogs worn or damaged. Rounding off of the dog edges and the mating recesses in the adjacent pinion can lead to jumping out of gear when under load. The gears should be inspected and renewed. Attempting to reprofile the dogs is not recommended.
● Selector forks, gearchange drum and pinion grooves worn. Extreme wear of these interconnected items can occur after high mileages especially when lubrication has been neglected. The worn components must be renewed.
● Gear pinions, bushes and shafts worn. Renew the worn components.
● Bent gearchange shaft. Often caused by dropping the machine on the gear lever.
● Gear pinion tooth broken. Chipped teeth are unlikely to cause jumping out of gear once the gear has been selected fully; a tooth which is completely broken off, however, may cause problems in this respect and in any event will cause transmission noise.

26 Overselection

● Pawl spring weak or broken. Renew the spring.
● Detent stopper arm or cam worn or broken. Renew the damaged item.
● Stopper arm spring worn or broken. Renew the spring.

Abnormal engine noise.

27 Knocking or pinking

● See Section 16.

28 Piston slap or rattling from cylinder

● Cylinder bore/piston clearance excessive. Resulting from wear or partial seizure. This condition can often be heard as a high, rapid tapping noise when the engine is under little or no load, particularly when power is just beginning to be applied. Reboring to the next correct oversize should be carried out and a new oversize piston fitted.
● Connecting rod bent. This can be caused by over-revving, trying to start a very badly flooded engine (resulting in a hydraulic lock in the cylinder) or by earlier mechanical failure. Attempts at straightening a bent connecting rod are not recommended. Careful inspection of the crankshaft should be made before renewing the damaged connecting rod.
● Gudgeon pin, piston boss bore or small-end bearing wear or seizure. Excess clearance or partial seizure between normal moving parts of

these items can cause continuous or intermittent tapping noises. Rapid wear or seizure is caused by lubrication starvation.
● Piston rings worn, broken or sticking. Renew the rings after careful inspection of the piston and bore.

29 Other noises

● Big-end bearing wear. A pronounced knock from within the crankcase which worsens rapidly is indicative of big-end bearing failure as a result of extreme normal wear or lubrication failure. Remedial action in the form of a bottom end overhaul should be taken; continuing to run the engine will lead to further damage including the possibility of connecting rod breakage.
● Main bearing failure. Extreme normal wear or failure of the main bearings is characteristically accompanied by a rumble from the crankcase and vibration felt through the frame and footrests. Renew the worn bearings and carry out a very careful examination of the crankshaft.
● Crankshaft excessively out of true. A bent crank may result from over-revving or damage from an upper cylinder component or gearbox failure. Damage can also result from dropping the machine on either crankshaft end. Straightening of the crankshaft is not possible in normal circumstances; a replacement item should be fitted.
● Engine mounting loose. Tighten all the engine mounting nuts and bolts.
● Cylinder head gasket leaking. The noise most often associated with a leaking head gasket is a high pitched squeaking, although any other noise consistent with gas being forced out under pressure from a small orifice can also be emitted. Gasket leakage is often accompanied by oil seepage from around the mating joint or from the cylinder head holding down bolts and nuts. Leakage results from insufficient or uneven tightening of the cylinder head fasteners, or from random mechanical failure. Retightening to the correct torque figure will, at best, only provide a temporary cure. The gasket should be renewed at the earliest opportunity.
● Exhaust system leakage. Popping or crackling in the exhaust system, particularly when it occurs with the engine on the overrun, indicates a poor joint either at the cylinder port or at the exhaust pipe/silencer connection. Failure of the gasket or looseness of the clamp should be looked for.

Abnormal transmission noise

30 Clutch noise

● Clutch outer drum/friction plate tang clearance excessive.
● Clutch outer drum/spacer clearance excessive.
● Clutch outer drum/thrust washer clearance excessive.
● Primary drive gear teeth worn or damaged.

31 Transmission noise

● Bearing or bushes worn or damaged. Renew the affected components.
● Gear pinions worn or chipped. Renew the gear pinions.
● Metal chips jammed in gear teeth. This can occur when pieces of metal from any failed component are picked up by a meshing pinion. The condition will lead to rapid bearing wear or early gear failure.
● Gearbox oil level too low. Top up immediately to prevent damage to gearbox and engine.
● Gearchange mechanism worn or damaged. Wear or failure of certain items in the selection and change components can induce mis-selection of gears (see Section 24) where incipient engagement of more than one gear set is promoted. Remedial action, by the overhaul of the gearbox, should be taken without delay.
● Chain snagging on cases or cycle parts. A badly worn chain or one that is excessively loose may snag or smack against adjacent components.

Exhaust smokes excessively

32   White/blue smoke (caused by oil burning)

● Oil pump settings incorrect. Check and reset the oil pump as described in Chapter 2.
● Crankshaft main bearing oil seals worn. Wear in the main bearing oil seals, often in conjunction with wear in the bearings themselves, can allow transmission oil to find its way into the crankcase and thence to the combustion chamber. This condition is often indicated by a mysterious drop in the transmission oil level with no sign of external leakage.
● Accumulated oil deposits in exhaust system. If the machine is used for short journeys only it is possible for the oil residue in the exhaust gases to condense in the relatively cool silencer. If the machine is then taken for a longer run in hot weather, the accumulated oil will burn off producing ominous smoke from the exhaust.

33   Black smoke (caused by over-rich mixture)

● Air filter element clogged. Clean or renew the element.
● Main jet loose or too large. Remove the float chamber to check for tightness of the jet. If the machine is used at high altitudes rejetting will be required to compensate for the lower atmospheric pressure.
● Cold start mechanism jammed on. Check that the mechanism works smoothly and correctly.
● Fuel level too high. The fuel level is controlled by the float height which can increase as a result of wear or damage. Remove the float bowl and check the float height. Check also that the floats have not punctured; a punctured float will lose buoyancy and allow an increased fuel level.
● Float valve needle stuck open. Caused by dirt or a worn valve. Clean the float chamber or renew the needle and valve seat.

Poor handling or roadholding

34   Directional instability

● Steering head bearing adjustment too tight. This will cause rolling or weaving at low speeds. Re-adjust the bearings.
● Steering head bearings worn or damaged. Correct adjustment of the bearings will prove impossible to achieve if wear or damage has occurred. Inconsistent handling will occur including rolling or weaving at low speed and poor directional control at indeterminate higher speeds. The steering head bearings should be dismantled for inspection and renewed if required. Lubrication should also be carried out.
● Bearing races pitted or dented. Impact damage caused, perhaps, by an accident or riding over a pot-hole can cause indentation of the bearing, usually in one position. This should be noted as notchiness when the handlebars are turned. Renew and lubricate the bearings.
● Steering stem bent. This will occur only if the machine is subjected to a high impact such as hitting a curb or a pot-hole. The lower yoke/stem should be renewed; do not attempt to straighten the stem.
● Front or rear tyre pressures too low.
● Front or rear tyre worn. General instability, high speed wobbles and skipping over white lines indicates that tyre renewal may be required. Tyre induced problems, in some machine /tyre combinations, can occur even when the tyre in question is by no means fully worn.
● Swinging arm bushes worn. Difficulties in holding line, particularly when cornering or when changing power settings indicates wear in the swinging arm bushes. The swinging arm should be removed from the machine and the bushes renewed.
● Swinging arm flexing. The symptoms given in the preceding paragraph will also occur if the swinging arm fork flexes badly. This can be caused by structural weakness as a result of corrosion, fatigue or impact damage, or because the rear wheel spindle is slack.
● Wheel bearings worn. Renew the worn bearings.
● Loose wheel spokes. The spokes should be tightened evenly to maintain tension and trueness of the rim.
● Tyres unsuitable for machine. Not all available tyres will suit the characteristics of the frame and suspension, indeed, some tyres or tyre

combinations may cause a transformation in the handling characteristics. If handling problems occur immediately after changing to a new tyre type or make, revert to the original tyres to see whether an improvement can be noted. In some instances a change to what are, in fact, suitable tyres may give rise to handling deficiences. In this case a thorough check should be made of all frame and suspension items which affect stability.

35   Steering bias to left or right

● Rear wheel out of alignment. Caused by uneven adjustment of chain tensioner adjusters allowing the wheel to be askew in the fork ends. A bent rear wheel spindle will also misalign the wheel in the swinging arm.
● Wheels out of alignment. This can be caused by impact damage to the frame, swinging arm, wheel spindles or front forks. Although occasionally a result of material failure or corrosion it is usually as a result of a crash.
● Front forks twisted in the steering yokes. A light impact, for instance with a pot-hole or low curb, can twist the fork legs in the steering yokes without causing structural damage to the fork legs or the yokes themselves. Re-alignment can be made by loosening the yoke pinch bolts and the wheel spindle. Re-align the wheel with the handlebars and tighten the bolts working upwards from the wheel spindle. This action should be carried out only when there is no chance that structural damage has occurred.

36   Handlebar vibrates or oscillates

● Tyres worn or out of balance. Either condition, particularly in the front tyre, will promote shaking of the fork assembly and thus the handlebars. A sudden onset of shaking can result if a balance weight is displaced during use.
● Tyres badly positioned on the wheel rims. A moulded line on each wall of a tyre is provided to allow visual verification that the tyre is correctly positioned on the rim. A check can be made by rotating the tyre; any misalignment will be immediately obvious.
● Wheel rims warped or damaged. Inspect the wheels for runout as described in Chapter 5.
● Swinging arm bushes worn. Renew the bushes.
● Wheel bearings worn. Renew the bearings.
● Steering head bearings incorrectly adjusted. Vibration is more likely to result from bearings which are too loose rather than too tight. Re-adjust the bearings.
● Loose fork component fasteners. Loose nuts and bolts holding the fork legs, wheel spindle, mudguards or steering stem can promote shaking at the handlebars. Fasteners on running gear such as the forks and suspension should be check tightened occasionally to prevent dangerous looseness of components occurring.
● Engine mounting bolts loose. Tighten all fasteners.

37   Poor front fork performance

● Damping fluid level incorrect. If the fluid level is too low poor suspension control will occur resulting in a general impairment of roadholding and early loss of tyre adhesion when cornering and braking. Too much oil is unlikely to change the fork characteristics unless severe overfilling occurs when the fork action will become stiffer and oil seal failure may occur.
● Damping oil viscosity incorrect. The damping action of the fork is directly related to the viscosity of the damping oil. The lighter the oil used, the less will be the damping action imparted. For general use, use the recommended viscosity of oil, changing to a slightly higher or heavier oil only when a change in damping characteristic is required. Overworked oil, or oil contaminated with water which has found its way past the seals, should be renewed to restore the correct damping performance and to prevent bottoming of the forks.
● Damping components worn or corroded. Advanced normal wear of

the fork internals is unlikely to occur until a very high mileage has been covered. Continual use of the machine with damaged oil seals which allows the ingress of water, or neglect, will lead to rapid corrosion and wear. Dismantle the forks for inspection and overhaul.

● Weak fork springs. Progressive fatigue of the fork springs, resulting in a reduced spring free length, will occur after extensive use. This condition will promote excessive fork dive under braking, and in its advanced form will reduce the at-rest extended length of the forks and thus the fork geometry. Renewal of the springs as a pair is the only satisfactory course of action.

● Bent stanchions or corroded stanchions. Both conditions will prevent correct telescoping of the fork legs, and in an advanced state can cause sticking of the fork in one position. In a mild form corrosion will cause stiction of the fork thereby increasing the time the suspension takes to react to an uneven road surface. Bent fork stanchions should be attended to immediately because they indicate that impact damage has occurred, and there is a danger that the forks will fail with disastrous consequences.

38 Front fork judder when braking (see also Section 46)

● Wear between the fork stanchions and the fork legs. Renewal of the affected components is required.

● Slack steering head bearings. Re-adjust the bearings.

● Warped brake drum. If irregular braking action occurs fork judder can be induced in what are normally serviceable forks. Renew the damaged brake components.

39 Poor rear suspension performance

● Rear suspension unit damper worn out or leaking. The damping performance of most rear suspension units falls off with age. This is a gradual process, and thus may not be immediately obvious. Indications of poor damping include hopping of the rear end when cornering or braking, and a general loss of positive stability.

● Weak rear spring. If the suspension unit spring fatigues it will promote excessive pitching of the machine and reduce the ground clearance when cornering. As replacement springs are not available separately the complete suspension unit will require renewal.

● Swinging arm flexing or bushes worn. See Sections 34 and 36.

● Bent suspension unit damper rod. This is likely to occur only if the machine is dropped or if seizure of the piston occurs. If either happens the suspension unit should be renewed.

● Worn suspension linkage pivot bearings. Overhaul the assembly as described in Chapter 4.

Abnormal frame and suspension noise

40 Front end noise

● Oil level low or too thin. This can cause a 'spurting' sound and is usually accompanied by irregular fork action.

● Spring weak or broken. Makes a clicking or scraping sound. Fork oil will have a lot of metal particles in it.

● Steering head bearings loose or damaged. Clicks when braking. Check, adjust or renew.

● Fork clamps loose. Make sure all fork clamp pinch bolts are tight.

● Fork stanchion bent. Good possibility if machine has been dropped. Repair or renew stanchion.

41 Rear suspension noise

● Fluid level too low. Leakage of a suspension unit, usually evident by oil on the outer surfaces, can cause a spurting noise. The suspension unit should be renewed.

● Defective rear suspension unit with internal damage. Renew the suspension unit.

Brake problems

42 Brakes are spongy or ineffective

● Brake cable deterioration (front brake). Damage to the outer cable by stretching or being trapped will give a spongy feel to the brake lever. The cable should be renewed. A cable which has become corroded due to old age or neglect of lubrication will partially seize, making operation very heavy. Lubrication at this stage may overcome the problem but the fitting of a new cable is recommended.

● Worn brake linings. Determine lining wear using the external brake wear indicator mark on the cam and backplate, or by removing the wheel and withdrawing the brake backplate. Renew the shoes as a pair if the linings are worn below the recommended limit.

● Worn brake camshaft. Wear between the camshaft and the bearing surface will reduce brake feel and reduce operating efficiency. Renewal of one or both items will be required to rectify the fault.

● Worn brake cam and shoe ends. Renew the worn components.

● Linings contaminated with dust or grease. Any accumulations of dust should be cleaned from the brake assembly and drum using a petrol dampened cloth. Do not blow or brush off the dust because it is asbestos based and thus harmful if inhaled. Light contamination from grease can be removed from the surface of the brake linings using a solvent; attempts at removing heavier contamination are less likely to be successful because some of the lubricant will have been absorbed by the lining material which will severely reduce the braking performance.

43 Brake drag

● Incorrect adjustment. Re-adjust the brake operating mechanism.

● Drum warped or oval. This can result from overheating or impact or uneven tension of the wheel spokes. The condition is difficult to correct, although if slight ovality only occurs, skimming the surface of the brake drum can provide a cure. This is work for a specialist engineer. Renewal of the complete wheel hub is normally the only satisfactory solution.

● Weak brake shoe return springs. This will prevent the brake lining/shoe units from pulling away from the drum surface once the brake is released. The springs should be renewed.

● Brake camshaft, lever/pedal pivot or cable poorly lubricated. Failure to attend to regular lubrication of these areas will increase operating resistance which, when compounded, may cause tardy operation and poor release movement.

44 Brake lever or pedal pulsates in operation

● Drums warped or oval. This can result from overheating or impact or uneven spoke tension. This condition is difficult to correct, although if slight ovality only occurs skimming the surface of the drum can provide a cure. This is work for a specialist engineer. Renewal of the hub is normally the only satisfactory solution.

45 Brake noise

● Drum warped or oval. This can cause intermittent rubbing of the brake linings against the drum. See the preceding Section.

● Brake linings glazed. This condition, usually accompanied by heavy lining dust contamination, often induces brake squeal. The surface of the linings may be roughened using glass-paper or a fine file.

46 Brake induced fork judder

● Worn front fork stanchions and legs, or worn or badly adjusted steering head bearings. These conditions, combined with uneven or pulsating braking as described in Section 44 will induce more or less judder when the brakes are applied, dependent on the degree of wear and poor brake operation. Attention should be given to both areas of malfunction. See the relevant Sections.

Electrical problems

47   Battery dead or weak

● Battery faulty. Battery life should not be expected to exceed 3 to 4 years. Gradual sulphation of the plates and sediment deposits will reduce the battery performance. Plate and insulator damage can often occur as a result of vibration. Complete power failure, or intermittent failure, may be due to a broken battery terminal. Lack of electrolyte will prevent the battery maintaining charge.

● Battery leads making poor contact. Remove the battery leads and clean them and the terminals, removing all traces of corrosion and tarnish. Reconnect the leads and apply a coating of petroleum jelly to the terminals.

● Load excessive. If additional items such as spot lamps, are fitted, which increase the total electrical load above the maximum alternator output, the battery will fail to maintain full charge. Reduce the electrical load to suit the electrical capacity.

● Rectifier failure.

● Flywheel generator coils open-circuit or shorted.

● Charging circuit shorting or open circuit. This may be caused by frayed or broken wiring, dirty connectors or a faulty ignition switch. The system should be tested in a logical manner. See Section 50.

48   Battery overcharged

● Regulator faulty. Overcharging is indicated if the battery becomes hot or it is noticed that the electrolyte level falls repeatedly between checks. In extreme cases the battery will boil causing corrosive gases and electrolyte to be emitted through the vent pipes.

● Battery wrongly matched to the electrical circuit. Ensure that the specified battery is fitted to the machine.

49   Total electrical failure

● Fuse blown. Check the main fuse. If a fault has occurred, it must be rectified before a new fuse is fitted.

● Battery faulty. See Section 47.

● Earth failure. Check that the frame main earth strap from the black/white earth circuit is securely affixed to the frame and is making a good contact.

● Ignition switch or power circuit failure. Check for current flow through the battery positive lead (red) to the ignition switch. Check the ignition switch for continuity.

50   Circuit failure

● Cable failure. Refer to the machine's wiring diagram and check the circuit for continuity. Open circuits are a result of loose or corroded connections, either at terminals or in-line connectors, or because of broken wires. Occasionally, the core of a wire will break without there being any apparent damage to the outer plastic cover.

● Switch failure. All switches may be checked for continuity in each switch position, after referring to the switch position boxes incorporated in the wiring diagram for the machine. Switch failure may be a result of mechanical breakage, corrosion or water.

● Fuse blown. Renew the fuse, if blown, only after the fault has been identified and rectified.

51   Bulbs blowing repeatedly

● Vibration failure. This is often an inherent fault related to the natural vibration characteristics of the engine and frame and is, thus, difficult to resolve. Modifications of the lamp mounting, to change the damping characteristics, may help.

● Intermittent earth. Repeated failure of one bulb, particularly where the bulb is fed directly from the generator, indicates that a poor earth exists somewhere in the circuit. Check that a good contact is available at each earthing point in the circuit.

● Reduced voltage. Do not overload the system with additional electrical equipment in excess of the system's power capacity and ensure that all circuit connections are maintained clean and tight.

Routine maintenance

Engine/gearbox
Spark plug:
    Type.. NGK BP6ES or ND W20EP
    Gap.. 0.6 – 0.8 mm (0.024 – 0.032 in)
Ignition timing.. $18 \pm 2°$ BTDC @6000 rpm
Idle speed.. 1400 ± 100 rpm
Throttle cable free play at twistgrip .... 0.5 – 1.0 mm (0.02 – 0.04 in)
Clutch cable free play at lever.. 4 mm (0.16 in)

Cycle parts
Brakes:
    Front brake cable free play at lever................................... 5 mm (0.2 in)
    Rear brake pedal free play... 20 – 30 mm (0.8 – 1.2 in)
Drive chain free play .... 30 – 40 mm (1.2 – 1.6 in)
Tyre pressures – cold:
    Front .... 1.5 kg/cm² (20 psi)
    Rear... Solo – 1.75 kg/cm² (24 psi), Pillion – 2.0 kg/cm² (28 psi)
Tyre tread minimum depth – manufacturer's recommendation......... 3 mm (0.04 in)

Recommended lubricants and fluids
Engine:
    Fuel grade.. Unleaded or low-lead, minimum 85 octane (RON/RM)
    Recommended oil.. Suzuki CCI, CCI Super or a good quality two-stroke oil suitable for oil injection systems

Transmission:
    Capacity – at oil change... 850 cc (1.5 Imp pint)
    Recommended oil.. SAE 20W/40 motor oil
Air filter.. SAE 20W/40 motor oil
Front forks:
    Capacity per leg.. 185 cc (6.5 fl oz)
    Recommended oil.. SAE 10W fork oil
Final drive chain .... Aerosol chain lubricant
Wheel bearings, brake camshafts and speedometer drive gearbox .. High melting-point grease
Steering head bearings.. Good quality general purpose grease
Swinging arm and suspension linkage pivots .... Good quality general purpose grease
All control pivots, instrument cables, stand pivots and throttle
twistgrip .... Good quality general purpose grease
Control cables .... Engine oil

Introduction

Periodic routine maintenance is a continuous process which should commence immediately the machine is used. The object is to maintain all adjustments and to diagnose and rectify minor defects before they develop into more extensive, and often more expensive, problems.

It follows that if the machine is maintained properly, it will both run and perform with optimum efficiency, and be less prone to unexpected breakdowns. Regular inspection of the machine will show up any parts which are wearing, and with a little experience, it is possible to obtain the maximum life from any one component, renewing it when it becomes so worn that it is liable to fail.

Regular cleaning can be considered as important as mechanical maintenance. This will ensure that all the cycle parts are inspected regularly and are kept free from accumulations of road dirt and grime.

Cleaning is especially important during the winter months, despite its appearance of being a thankless task which very soon seems pointless. On the contrary, it is during these months that the paintwork, chromium plating, and the alloy casings suffer the ravages of abrasive grit, rain and road salt. A couple of hours spent weekly on cleaning the machine will maintain its appearance and value, and highlight small points, like chipped paint, before they become a serious problem.

The various maintenance tasks are described under their respective mileage and calendar headings, and are accompanied by line drawings and photographs where pertinent.

It should be noted that the intervals between each maintenance task serve only as a guide. As the machine gets older, or if it is used under particularly arduous conditions, it is advisable to reduce the period between each check.

For ease of reference, most service operations are described in detail under the relevant heading. However, if further general information is required, this can be found under the pertinent Section heading and Chapter in the main text.

Apart from a good selection of workshop tools, a suitable stand which will support the machine and keep it upright with enough height to remove either wheel, is considered to be a wise investment. A strong wooden box is normally recommended but this is not always convenient for maintenance tasks. Metal 'paddock' type stands seem very popular and are frequently advertised in the motorcycle press.

Daily (pre-ride) checks

Before taking the machine out on the road there are certain checks which should be completed to ensure that it is in a safe and legal condition to be used.

1   Engine oil level

Although the machine is fitted with an engine oil level warning light, this serves as an emergency measure to show that there is only a small amount of oil left in the reservoir tank. Ideally, the level should be maintained to within an inch of the top of the filler tank neck. Remove the seat to gain access to the oil tank filler cap. Note that the ignition switch is provided with an extra position (C) which automatically switches on the oil level warning lamp so that the rider can be certain that the warning lamp circuit is functioning correctly. On switching the ignition to the On position, the lamp should extinguish, and will only light again when the oil level drops to the low position.

Use only a good quality two-stroke oil designed for motorcycle injection systems when topping up the tank.

2   Petrol level

Checking the petrol level may seem obvious, but it is all too easy to forget. Ensure that you have enough petrol to complete your journey, or at least to get you to the nearest filling station.

3   Brakes

Check that the front and rear brakes work effectively and without binding. Ensure that the rod linkages and the cables, as applicable, are lubricated and properly adjusted.

4   Tyres

Check the tyre pressures with a gauge that is known to be accurate. It is well worth purchasing a pocket gauge for this purpose because garage forecourt airlines are notoriously inaccurate. The pressures should be checked with the tyres *cold*.

At the same time as the tyre pressures are checked, examine the tyres themselves. Check them for damage, especially splitting of the sidewalls. Remove any small stones or other debris caught between the treads. This is particularly important on the rear tyre, where rapid deflation due to penetration of the inner tube will almost certainly cause total loss of control. When checking the tyres for damage, they should be examined for tread depth in view of both the legal and safety aspects. It is vital to keep the tread depth within the UK legal limits of 1 mm of depth over three-quarters of the tread breadth around the entire circumference. Note that new laws mean that no bald patches are now permissible on any part of the tread. Many riders, however, consider nearer 2 mm to be the limit for secure roadholding, traction, and braking, especially in adverse weather conditions. Note the manufacturer's minimum tread depth given in the Specifications.

5   Final drive chain

Check that the final drive chain is correctly adjusted and well lubricated. Refer to the Two-monthly/600 mile service interval if attention is required. Remember that if used in adverse weather conditions, the chain will require more frequent even daily lubrication.

6   Cables, controls and steering

Check the throttle twistgrip, clutch and gear levers to ensure that they are securely fastened and working properly. If a bolt is going to work loose or a cable snap, it is better that it does so with the machine at a standstill than when riding. Check the steering for signs of excessive play or restriction of movement.

7   Lights and speedometer

Check that all lights, turn signals, horn and speedometer are working correctly to make sure that the machine complies with all legal requirements in this respect.

Two-monthly or every 600 miles (1000 km)

This is where the proper procedure of routine maintenance begins. The daily checks serve to ensure that the machine is safe and legal to use but will contribute little to maintenance other than to give the owner an accurate picture of what item needs attention. In addition, they will give early warning of any faults which are about to appear. When performing the two-monthly maintenance tasks, carry out the daily checks first.

1   Transmission oil

Check the amount of oil in the transmission by removing the level screw in the right-hand engine casing with the machine standing upright and on level ground. Oil should slowly trickle out of the screw orifice if the level is correct. If necessary add oil of the specified type via the filler plug. Refit and tighten the level screw having checked the condition of its sealing washer.

2   Brake adjustment

If the front brake adjustment is correct, the amount of free play from the butt end of the handlebar lever to its clamp should be 5 mm (0.2 in). Minor adjustments can be made to the front brake after pulling back the rubber cover from the handlebar lever, slackening the lock ring and turning the adjuster until the correct free play is obtained.

If more adjustment is necessary, slacken the lock ring once again and screw the adjuster fully into the handlebar clamp. Loosen the brake cable clamp from the top of the fork lower leg to allow cable movement.

Maintain engine oil level to within an inch of filler neck

Check tyre pressures with an accurate gauge

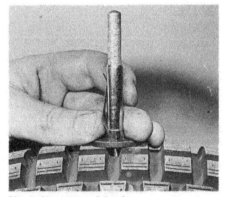
Check the tyre tread depth

Transmission oil level is checked via level screw in crankcase right-hand cover

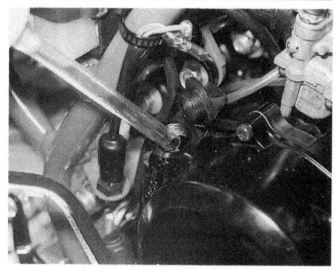

Remove filler plug and top up transmission oil if required

Slacken the two locknuts at the cable lower end and take up the adjustment until the correct amount of cable free play is given at the handlebar lever.

Spin the front wheel to ensure that the brake shoes are not binding and that the wheel is free to rotate, then apply the front brake hard to settle the cable and brake components. Check that the adjustment has not altered; reset if required. Remember to fully tighten all locknuts and clamps, and secure the rubber cable protecting sleeves.

Check the front brake light switch after making any adjustments to the front brake. The rear stop lamp should illuminate just as pressure is felt at the brake lever. The switch is located beneath the lever and adjustment is made by loosening its fitting screw and moving the switch body up or down in the appropriate direction.

Rear brake adjustment is by means of a single nut at the end of the brake operating rod. Prior to the brake adjustment, adjust the brake pedal height to a position which feels comfortable to use; the standard setting being 10 mm (0.4 in) below the level of the footrest. Brake pedal free play should be between 20 – 30 mm (0.8 – 1.2 in) travel at the pedal tip; if adjustment is required turn the nut at the end of the operating rod. Check that the rear wheel rotates easily and that the stop lamp switch is functioning correctly. Remember that the stop lamp switch height must be checked, and if necessary, adjusted every time the rear brake adjustment has been altered. To adjust the switch, turn its plastic sleeve nut as required until the stop lamp bulb lights after the free play has been taken up by the brake pedal and the brake just comes into operation.

Lower adjuster is used to take up large amounts of front brake cable free play

Slacken brake cable clamp on fork leg whilst adjustment is made

Use adjuster shown to alter brake pedal height

Rear brake adjustment is made using nut at end of brake rod

Rear brake stop lamp switch height may require adjusting after rear brake adjustment

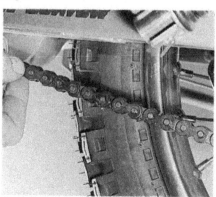

Chain tension is checked midway between the two sprockets

Complete brake maintenance by oiling all lever pivot points, all exposed cable ends and the rear brake linkage with a few drops of oil from a can. Do not apply an excessive amount of oil to the linkages in case any surplus should find its way into the brake drum or onto the tyre.

A regular check must also be made to ensure that the friction material on the brake shoes has not worn past its service limit and to ensure that worn items are renewed in order to maintain peak brake efficiency. This can be checked without direct inspection of the shoes. An extension of the groove cut into the brake camshaft must be within the limit of the line cast on the brake backplate when the brake is correctly adjusted and fully applied. If the line is seen to be outside the arc on the backplate, the friction material is worn and the brake shoes must be renewed.

Overhauling the brakes is preceded by removing the wheel as described in the relevant section of Chapter 5. The brake components can then be dismantled, cleaned, checked and reassembled following instructions given in the same chapter.

3 Final drive chain lubrication and adjustment

For the purpose of lubrication, one of the many proprietary aerosol chain lubricants can be applied very conveniently. It should be used more frequently if the machine is used in wet or dusty conditions. Gear oil can be used but its effective life is lessened due to the non cling properties which it possesses.

Check the chain tension by raising the rear wheel off the ground and revolving it until the tightest spot along the length of chain is found. Measure chain free play at the tightest spot, midway between the front and rear sprockets on the chain lower run; this should be 30 – 40 mm (1.2 – 1.6 in).

If adjustment is necessary, loosen the wheel spindle and torque arm to brake backplate nuts. Tighten the chain by turning each adjuster nut at the ends of the swinging arm clockwise by an equal number of turns. Use the marks stamped in the sides of the swinging arm as a guide to wheel alignment. For a more accurate check position the machine on an improvised stand, so that it is upright, and lay a plank of wood (or draw a length of string) parallel to the machine and as high as possible so that it touches both walls of the rear tyre. Wheel alignment is correct when the plank or string is equidistant from both walls of the front tyre when tested on both sides of the machine, as shown in the accompanying illustration.

With adjustment correct, lightly tighten the spindle nut, spin the wheel and apply the rear brake hard; this will centre the brake backplate on the drum. Tighten the spindle nut to the correct torque setting, see Chapter 5, recheck chain adjustment and tighten the torque arm nut. Fit a new split pin and ensure that its ends are spread correctly.

Check that the wheel spins freely and if necessary adjust the brake and stop lamp switch height.

4 Cleaning and resetting the spark plug

Detach the spark plug cap and remove the spark plug. Clean the electrodes with a wire brush followed by a strip of fine emery cloth. Check the plug gap and reset if necessary to that specified. Make adjustments by bending the outer electrode only, never the inner (central) electrode. Before refitting the spark plug smear its threads with

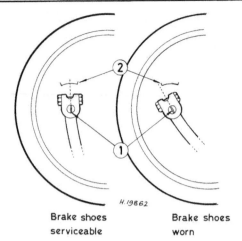

H.19862

Brake shoes
serviceable

Brake shoes
worn

External brake shoe wear indicator marks

1   Camshaft groove
2   Wear indicator arc

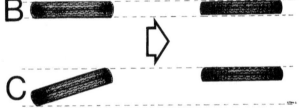

Wheel alignment check

A and C – incorrect
B – correct

a graphited grease; this will aid subsequent removal.

Note that the appearance of the spark plug can be used to assess the general condition of the engine.

5 Check the carburettor, throttle cable and fuel pipe

If rough running of the engine has developed, some adjustment of the carburettor pilot screw and tick-over speed may be required. If this

Make chain tension adjustment using adjuster nuts on swinging arm ends

Use stamped alignment marks to ensure each adjuster is moved by the same amount

Remove the spark plug, clean its electrodes and check the gap

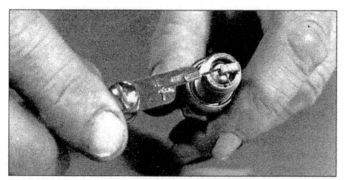

Electrode gap check - use a wire type gauge for best results

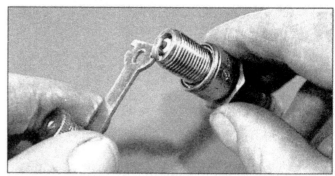

Electrode gap adjustment - bend the side electrode using the correct tool

Normal condition - A brown, tan or grey firing end indicates that the engine is in good condition and that the plug type is correct

Ash deposits - Light brown deposits encrusted on the electrodes and insulator, leading to misfire and hesitation. Caused by excessive amounts of oil in the combustion chamber or poor quality fuel/oil

Carbon fouling - Dry, black sooty deposits leading to misfire and weak spark. Caused by an over-rich fuel/air mixture, faulty choke operation or blocked air filter

Oil fouling - Wet oily deposits leading to misfire and weak spark. Caused by oil leakage past piston rings or valve guides (4-stroke engine), or excess lubricant (2-stroke engine)

Overheating - A blistered white insulator and glazed electrodes. Caused by ignition system fault, incorrect fuel, or cooling system fault

Worn plug - Worn electrodes will cause poor starting in damp or cold weather and will also waste fuel

Remove the drain plug to check for fuel contamination

is the case refer to Chapter 2, Section 7 for details. Do not make these adjustments unless they are obviously required, there is little to be gained by unwarranted attention to the carburettor. Complete carburettor maintenance by removing the drain plug on the float chamber, turning the petrol on, and allowing a small amount of fuel to drain through, thus flushing any water or dirt from the carburettor. Refit the drain plug securely and switch the petrol off.

Once the carburettor has been checked and reset if necessary, the throttle cable free play can be checked. Open and close the throttle several times, allowing it to snap shut under its own pressure. Ensure that it is able to shut off quickly and fully at all handlebar positions. Check that there is 0.5 – 1.0 mm (0.02 – 0.04 in) free play measured in terms of twistgrip rotation. If not, use the adjuster at the twistgrip to achieve the correct setting, completing the operation, if necessary, with the adjuster on the carburettor top. Open and close the throttle again to settle the cable and to check that adjustment is not disturbed.

Give the pipe which connects the fuel tap and carburettor a close visual examination, checking for cracks or any signs of leakage. In time, the synthetic rubber pipe will tend to deteriorate, and will eventually leak. Apart from the obvious fire risk, the leaking fuel will affect fuel economy. If the pipe is to be renewed, always use the correct replacement type to ensure a good leak-proof fit. Never use natural rubber tubing because this will tend to break up when in contact with petrol and will obstruct the carburettor jets.

6 Oil pump adjustment

As the oil pump is linked directly to the throttle cable, adjustments should only be carried out after the carburettor has been checked as described in the previous section.

The upper cylindrical portion of the carburettor body has an inspection aperture in its left-hand side, blanked off by a crosshead screw. Remove the screw then remove the left-hand crankcase cover which is

retained by four screws. Note that it should be possible to manoeuvre the cover to one side, obviating the need to disconnect the clutch cable. Slacken and remove the screws securing the oil pump cover and lift away the cover.

Open the throttle carefully and slowly by rotating the twistgrip until the small punch mark in the throttle slide comes into view. Carry on opening the throttle until the upper edge of the punch mark aligns exactly with the upper edge of the inspection aperture. Hold the twistgrip steady at this point and inspect the oil pump control lever alignment marks. The long scribed mark on the pump control lever should align exactly with the fixed mark on the pump body. If the two marks do not align, slacken the locknut on the oil pump cable adjuster set in the crankcase and rotate the adjuster as necessary until the two marks are exactly in line. Tighten the locknut and open and close the throttle several times to settle the cables. Check the setting again, altering it if necessary. Once the oil pump is correctly set, apply a few drops of oil to the oil pump cable and to the pump control lever pivot.

Carefully inspect the oil feed and delivery pipes for signs of leakage, bends or twisting which might cause an intermittent supply of oil. Any signs of leakage or damage should be attended to immediately and the system bled as described in Chapter 2, Section 14.

When the oil pump, cable, and oil lines have been inspected and re-set and lubricated as necessary, refit the pump cover, the crankcase left-hand cover, and slide the rubber cover back over the cable adjuster. Check that its sealing washer is in good condition and refit the inspection screw to the carburettor body.

7 Check the clutch adjustment

While the clutch can be adjusted at two points, at the operating mechanism and at the cable itself, it will suffice for the purposes of Routine maintenance to regard the operating mechanism as set and to make all normal adjustments using the cable length adjusters. The clutch is adjusted correctly when there is 4 mm (0.16 in) of free play in the cable, the free play being measured between the butt end of the clutch handlebar lever and its handlebar clamp.

Reset the cable free play, if necessary, using the adjusters provided. Use the lower cable adjuster first, reserving the handlebar adjuster for minor alterations.

In the event that all cable adjustment has been taken up using the cable adjusters, remove the small inspection plate which is retained to the left-hand crankcase cover by two screws. Loosen the locknuts and screw in fully the cable adjusters to give maximum cable free play. Slacken the nut at the centre of the clutch release arm and screw in the centre adjuster until light resistance is felt, then back off the adjuster by $\frac{1}{4} - \frac{1}{2}$ turn and tighten the locknut. The correct cable free play can once again be taken up on the adjusters. When the correct free play is set at the handlebar lever, tighten the cable adjuster locknuts.

Complete clutch maintenance by applying a few drops of oil to the lever pivots, adjuster threads and to all exposed lengths of inner cable.

8 Checking the battery

Note: Read the notes in Safety first before handling the battery.

The electrolyte level must be maintained between the level marks on the battery casing. Top up if necessary with distilled water; do not use tap water.

To view throttle valve punch mark remove inspection screw (arrowed)

Remove cover to gain access to oil pump

Pump control lever mark should align with fixed mark on body with throttle in specified position

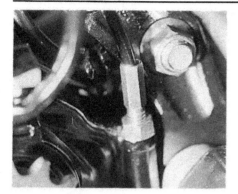

Make any adjustment required using cable adjuster at top of casing

Remove inspection cover to gain access to clutch release mechanism

Adjust mechanism via adjuster screw and locknut (arrowed)

The exposed battery terminals are prone to corroding, leading to poor electrical contact. Remove all traces of corrosion from the battery terminals using a scraper or emery cloth and remake the connections while the joint is still clean; smear each terminal with petroleum jelly (not grease) to prevent subsequent problems.

Finish off by checking that the vent pipe is secured to the battery and that it is correctly routed and free from kinks and blockages. Check also that the filler caps are tight.

9 Checking the steering and suspension

Check the operation of the front forks by pushing and pulling on the handlebars with the front brake applied. The forks should move smoothly, with no signs of stiffness or excessive play. Any knocking in the steering head may be indicative of worn or badly adjusted steering head bearings.

Raise the front wheel clear of the ground using a suitable support placed under the crankcase. Grasp the lower ends of the fork legs and push and pull them to check for play. It will be possible to feel any free movement in the steering head bearings. Turn the handlebars from lock to lock to check for roughness. If the above checks indicate wear in the steering head area, refer to Chapter 4 for details of renewal.

In order to adjust the steering head bearings, it will first be necessary to support the machine so that its front wheel is raised clear of the ground, to remove the ignition switch and handlebar clamps and to slacken the steering stem bolt. The upper and lower yoke clamp bolts will also have to be slackened in order to relax any pressures acting upon the steering head assembly and adjuster ring.

Bearing adjustment is correct when the adjuster ring is tightened, until resistance to movement is felt and then loosened $\frac{1}{4}$ to $\frac{1}{2}$ a turn. The adjuster ring should be rotated by means of a C-spanner.

Take great care not to overtighten the nut. It is possible to place a pressure of several tons on the head bearings by overtightening even

though the handlebars may seem to turn quite freely. Overtight bearings will cause the machine to roll at low speeds and give imprecise steering. Adjustment is correct if there is no play in the bearings and the handlebars swing to full lock either side when the machine is upright with the front wheel clear of the ground. Only a light tap on each end should cause the handlebars to swing. On completion of adjustment tighten the steering stem bolt, handlebar bolts and yoke clamp bolts to the specified torque setting.

Raise the rear wheel clear of the ground and check for play in the swinging arm pivot by pushing it from side to side. If any play is found, it will be necessary to dismantle and overhaul the swinging arm pivot assembly and to check for wear in the suspension linkage. This is described in detail in Chapter 4.

10 General checks

Proceed methodically around the machine, checking all nuts, bolts and screws, tightening any that may have worked loose. Be careful not to overtighten any component and make use of the torque wrench settings given in this Manual. Pay particular attention to the security of the engine mounting bolts, exhaust system fasteners and the cylinder head and barrel retaining nuts.

Six-monthly or every 2000 miles (3000 km)

First carry out the tasks listed under the previous time/mileage headings then proceed with the following:

1 Cleaning the air filter element

Engine wear and fuel consumption are greatly increased by lack of maintenance to the air filter element. This is the maximum interval for which the air filter must be removed and cleaned. Where the machine is used in particularly adverse conditions it is advised that cleaning takes place on a much more frequent basis.

Remove the left-hand sidepanel and the five screws which retain the filter cover. Remove the cover and separate the plastic frame from the foam element. Withdraw the element from the filter housing.

Examine the element carefully. It must be renewed if there are any splits or tears, or if the foam appears to be hardened through age. To clean the element, immerse it in a bath of non-flammable solvent such as white spirit. Petrol may be used, but be careful to take suitable precautions against the risk of fire. When the foam is clean, gently squeeze out the surplus solvent and allow the remainder to evaporate. Do not wring the element out, as this will damage the foam. Soak the cleaned, dry, element in the specified oil and gently squeeze out the surplus to leave the element slightly oily to the touch.

Filter reassembly is a straightforward reversal of the dismantling procedure, being careful to ensure that both the element and the filter cover are correctly seated so that no unfiltered air can bypass the element and enter the carburettor. A light application of grease to the sealing surfaces of the filter cover will help to achieve good sealing.

On no account should the air filter element be omitted while the engine is running in view of the increased noise level and of the high risk of severe damage to the engine due to overheating caused by the resultant weak mixture.

Maintain battery electrolyte level between level marks on casing

Air filter cover is retained by five screws   Withdraw filter element and supporting frame   Transmission oil drain plug location

2 Transmission oil change

Warm the engine/gearbox to normal operating temperature by taking the machine on a short journey. Stop the engine and position the machine upright on level ground with a container of at least 1 litre capacity placed underneath the gearbox drain plug. Wipe off any dirt from around the drain plug area, remove the oil filler and drain plugs into a container of approximately 2 pint capacity and allow the oil to drain fully (taking care to avoid scalding the hands). Examine the condition of the drain plug sealing gasket, renewing it if necessary. When the draining operation is complete refit the drain plug and tighten to the specified torque setting.

Refill the gearbox with the specified type and quantity of oil. Check that the level is correct by removing the oil level screw; with the machine positioned upright on its wheels the oil should slowly trickle out. Add or subtract oil until the level is correct. Refit the oil level screw and filler plug then wipe away any surplus oil.

3 Decarbonising the engine and exhaust system

Exhaust system

To carry out this operation, slacken and remove the exhaust mounting bolts and withdraw the exhaust system. Decarbonise the exhaust system following the procedure given in Chapter 2, Section 11.

Cylinder head, barrel and piston

Refer to Chapter 1 and remove the cylinder head and barrel.

Using a blunt-edged scraping tool to prevent scratching the soft alloy components, carefully remove all traces of carbon from the cylinder head combustion chamber, the piston crown and the barrel ports. Finish off using a soft rag and metal polish to give a smooth, polished finish to these areas which will reduce the ability of future carbon deposits to adhere so easily. Check that the piston rings are free in their grooves in the piston. If necessary, very carefully remove the rings using three thin strips of metal to free them, if they are stuck in place (See Fig. 1.7). Clean the skirt of the piston and remove any carbon in the ring grooves using a section of broken piston ring to ensure that the grooves are not damaged. Take this opportunity to inspect the bore surface, piston and rings for wear or damage; refer to Chapter 1 for further details.

Carefully clean the gasket surfaces of the crankcase, cylinder barrel, and cylinder head. Refit the cylinder barrel, and head as described in Chapter 1.

If the chain is not in need of renewal it should be removed from the machine and given a thorough clean and lubrication as described in Chapter 5, Section 10.

4 Final drive chain wear check and lubrication

Check for wear in the final drive chain following the instructions given in Chapter 5, Section 10. If the chain is worn and a new one is required, it is good practice to renew both sprockets at the same time. This is because the running together of new and worn components produces a far higher rate of wear than renewing the chain and sprockets as a set. Rear sprocket removal and refitting details are given in Chapter 5, Section 7. For removal and refitting of the gearbox sprocket refer to Chapter 1, Section 8 and Section 35 respectively.

If in serviceable condition, the chain should be removed from the machine and given a thorough clean and lubrication as described in Chapter 5. Section 10.

5 Check and lubricate the control cables

Check the outer cables for signs of damage, then examine the exposed portions of the inner cables. Any signs of kinking or fraying will indicate that renewal is required. To obtain maximum life and reliability from the cables they should be thoroughly lubricated. To do the job properly and quickly use one of the hydraulic cable oilers available from most motorcycle shops. Free one end of the cable and assemble the cable oiler as described by the manufacturer's instructions. Operate the oiler until oil emerges from the lower end, indicating that the cable is lubricated throughout its length. This process will expel any dirt or moisture and will prevent its subsequent ingress.

If a cable oiler is not available, an alternative is to remove the cable from the machine. Hang the cable upright and make up a small funnel arrangement using plasticine or by taping a plastic bag around the upper end. Fill the funnel with oil and leave it overnight to drain through. Note that where nylon-lined cables are fitted, they should be used dry or lubricated with a silicone-based lubricant suitable for this application. On no account use ordinary engine oil because this will cause the liner to swell, pinching the cable.

When refitting the cables, ensure that they are routed in easy curves and that full use is made of any guide or clamps that have been provided to secure the cable out of harm's way. Adjustment of the individual cables is described under previous Routine maintenance headings.

Check all pivots and control levers, cleaning and lubricating them to prevent wear or corrosion. Where necessary, dismantle and clean any moving part which may have become stiff in operation.

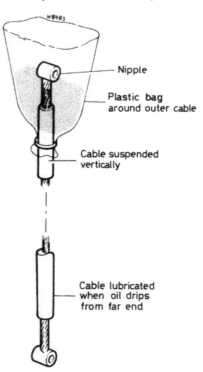

Oiling a control cable

Annually or every 4000 miles (6000 km)

Complete all the tasks listed under the previous mileage/time headings and then carry out the following:

1 Spark plug renewal

Remove and discard the spark plug regardless of its apparent condition. Although it may give acceptable service beyond this mileage limit, it is unlikely that it is still working at peak efficiency. Always fit the recommended type and heat range of spark plug and check that it is correctly gapped before fitting. Smear its threads with graphited grease to assist subsequent removal.

It is advisable to carry a spare spark plug with the machine's tool kit having first set the electrodes to the correct gap. Spark plug fouling is probably the most common cause for a two-stroke motorcycle to break down.

2 Clean the fuel tap sediment bowl

Switch the fuel tap to the 'Off' position and remove the sediment bowl from the bottom of the fuel tap by unscrewing it with a close-fitting ring spanner, then pick out the sealing O-ring. Remove all traces of dirt from the bowl and check the condition of the sealing O-ring, renewing it if it is worn or damaged. Refit the O-ring and the sediment bowl, tightening the bowl by just enough to nip the O-ring tight.

Note that if excessive traces of dirt or water appear in the sediment bowl, the fuel tank must be drained, removed from the machine and flushed out to prevent blockages in the fuel system. If draining of the fuel tank is necessary, the opportunity should be taken to dismantle the tap for cleaning, as described in Section 3 of Chapter 2, this being necessary to gain access to the gauze filter.

3 Cleaning the carburettor

Suzuki recommend that the carburettor be removed and dismantled for cleaning annually. This should only be necessary if there is real evidence of fuel contamination or if the machine's performance is suffering as a result. Full details of removing and dismantling the carburettor can be found in Chapter 2.

4 Change the fork oil

Regular oil changes will eliminate the possibility of deterioration of the fork oil. As each fork leg has to be removed, an excellent opportunity is given to examine the forks for wear. This operation also fits in well with other maintenance tasks required at this interval.

Remove the front wheel and fork legs from the machine as described in Chapters 5 and 4 respectively. Follow the procedure in Chapter 4, Section 3 for removal of the top plugs, spacers and springs, and draining of the fork oil. Information is also given on refilling the fork legs and measuring the oil level.

5 Grease the brake camshaft

The brake assemblies should be overhauled so that the brake camshaft can be cleaned and greased and the individual components examined. Remove the wheels as described in Chapter 5 and overhaul the brake assemblies as described in Sections 8 and 9 of the same Chapter.

6 Greasing the speedometer drive

With the front wheel removed from the machine, the speedometer drive components can be withdrawn and dismantled as described in Section 17 of Chapter 4. Lubricate each component with a smear of high melting-point grease before fitting and pack a small amount around the installed components to provide a supply of lubricant. Also check the oil seal fitted in the backplate; failure of this seal will allow grease to pass through to the brake linings.

7 Greasing the speedometer and tachometer cables

Unscrew their knurled rings and free each cable from its instrument and drive mechanism. Note that it will be necessary to remove the headlamp cowling to gain access to the instrument head cable connections. Release the cables from any ties provided, remove them from the machine and separate each inner cable from the outer. Look for signs of fraying or tight spots, indicated by shiny areas and inspect the outer cables for signs of fraying, kinking or any other damage. Renew either cable if necessary.

Smear a small quantity of grease to the inner cables, stopping about six inches from the top. This will prevent the instruments from becoming contaminated as the grease is pushed up by the action of the rotating cables. Insert the inner cables into their outers and refit the cables to the machine. Pay particular attention to the cable routing and make use of any cable ties or guides provided.

8 Greasing the suspension system

The manufacturer recommends that the suspension linkage and swinging arm components are examined and greased at this interval. This operation requires the dismantling of the respective assemblies and is given in detail in Chapter 4, Sections 8 to 10.

Two yearly or every 8000 miles (12 000 km)

Carry out the operations listed under the previous headings, then complete the following:

1 Greasing the steering head bearings

Remove the front wheel and both fork legs from the machine as described in Chapter 5 and 4 respectively, then dismantle and check the steering head bearings. Renew if worn, otherwise repack the bearings with fresh grease and readjust after assembly. See Chapter 4 for further information.

Additional routine maintenance

1 Renew the fuel feed pipe

Because the fuel feed pipe from the fuel tank to the carburettor is constructed of thin walled synthetic rubber, it will be affected by heat and the elements over a period of time. The manufacturer recommends that the pipe be renewed every four years irrespective of its apparent condition. Refer to Chapter 2, Section 4 for further information.

2 Cleaning the machine

Keeping the motorcycle clean should be considered as an important part of the routine maintenance, to be carried out whenever the need arises. A machine cleaned regularly will not only succumb less speedily to the inevitable corrosion of external surfaces, and hence maintain its market value, but will be far more approachable when the time comes for maintenance or service work. Furthermore, loose or failing components are more readily spotted when not partially obscured by a mantle of road grime and oil.

Surface dirt should be removed using a sponge and warm, soapy water; the latter being applied copiously to remove the particles of grit which might otherwise cause damage to the paintwork and polished surfaces.

Oil and grease is removed most easily by the application of a cleaning solvent such as 'Gunk' or 'Jizer'. The solvent should be applied when the parts are still dry and worked in with a stiff brush. Large quantities of water should be used when rinsing off, taking care that water does not enter the carburettor, air filter or electrics.

If desired a polish such as Solvol Autosol can be applied to the aluminium alloy parts to restore the original lustre. This does not apply in instances, much favoured by Japanese manufacturers, where the components are lacquered. Application of a wax polish to the cycle parts and a good chrome cleaner to the chrome parts will also give a good finish. Always wipe the machine down if used in the wet, and make sure the chain is well lubricated. There is less chance of water getting into control cables if they are regularly lubricated, which will prevent stiffness of action.

Chapter 1 Engine, clutch and gearbox

Contents

Specifications

Engine

Type	Single cylinder, air-cooled, two-stroke
Capacity	49 cc (2.99 cu in)
Bore	41 mm (1.61 in)
Stroke	37.8 mm (1.49 in)
Compression ratio	7.1:1

Cylinder head

Gasket face maximum distortion	0.050 mm (0.0020 in)

Cylinder barrel

Standard bore	40.975 – 40.990 mm (1.6132 – 1.6138 in)
Service limit	41.065 mm (1.6167 in)
Piston/cylinder clearance	0.040 – 0.050 mm (0.0016 – 0.0020 in)
Service limit	0.120 mm (0.0047 in)
Gasket face maximum distortion (head to barrel)	0.050 mm (0.0020 in)

Piston

Standard piston OD	40.930 – 40.945 mm (1.6114 – 1.6120 in)

Service limit	40.855 mm (1.6085 in)	
Oversizes available	+ 0.5 mm, + 1 mm (+ 0.02 in, + 0.04 in)	
Gudgeon pin OD	11.994 – 12.000 mm (0.4722 – 0.4725 in)	
Service limit	11.980 mm (0.4717 in)	
Gudgeon pin bore in piston	11.998 – 12.006 mm (0.4724 – 0.4727 in)	
Service limit	12.030 mm (0.4736 in)	

Piston rings

	Riken	Teikoku
Manufacturer	Riken	Teikoku
Top surface marking	R	T
Type	Keystone	Keystone
Free end gap	4.5 mm (0.1772 in)	5.0 mm (0.1969 in)
Service limit	3.6 mm (0.1417 in)	4.0 mm (0.1575 in)
Fitted end gap	0.100 – 0.250 mm (0.0039 – 0.0098 in)	
Service limit	0.750 mm (0.0295 in)	
Piston ring/groove clearance	0.020 – 0.060 mm (0.0008 – 0.0024 in)	

Crankshaft

Connecting rod small-end ID	16.000 – 16.008 mm (0.6299 – 0.6302 in)
Service limit	16.040 mm (0.6315 in)
Maximum connecting rod side-to-side deflection (at small-end)	3.0 mm (0.1181 in)
Maximum crankshaft runout	0.05 mm (0.0020 in)
Width across flywheels	40.0 ± 0.1 mm (1.5748 ± 0.0039 in)

Primary drive

Type	Gear
Reduction ratio (no of teeth)	3.842:1 (73/19T)

Clutch

Type	Wet, multiplate
Number of plates:	
Plain	2
Friction	3
Friction plate:	
Thickness	4.4 – 4.6 mm (0.1732 – 0.1811 in)
Service limit	4.1 mm (0.1614 in)
Tongue width	11.8 - 12.0 mm (0.4646 - 0.4724 in)
Service limit	11.0 mm (0.4330 in)
Plain plate:	
Thickness	1.6 ± 0.1 mm (0.0630 ± 0.0039 in)
Maximum warpage	0.10 mm (0.0039 in)
Spring free length service limit	33.6 mm (1.3228 in)

Gearbox

Type	5 speed, constant mesh
Gear ratios (no of teeth):	
1st	3.166:1 (38/12T)
2nd	1.941:1 (33/17T)
3rd	1.380:1 (29/21T)
4th	1.083:1 (26/24T)
5th	0.923:1 (24/26T)
Selector fork claw thickness	4.25 – 4.35 mm (0.1673 – 0.1713 in)
Selector fork groove width	4.45 – 4.55 mm (0.1752 – 0.1791 in)
Selector fork/groove clearance	0.10 – 0.30 mm (0.0039 – 0.0118 in)
Service limit	0.50 mm (0.0197 in)
Input shaft length (1st to 2nd gear)	84.5 – 84.6 mm (3.3268 – 3.3307 in)

Final drive

Type	Chain and sprocket
Reduction ratio (no of teeth)	4.250:1 (51/12T)

Torque wrench settings

Component	kgf m	lbf ft
Cylinder head nut	0.8 – 1.2	6 – 9
Cylinder barrel nut	0.8 – 1.2	6 – 9
Clutch centre nut	4.0 – 6.0	29 – 43
Clutch spring bolts	0.8 – 1.2	6 – 9
Primary drive pinion nut	4.0 – 6.0	29 – 43
Generator rotor nut	4.0 – 6.0	29 – 43
Engine mouting bolts	3.7 – 4.5	27 – 33
Swinging arm pivot nut (engine rear mounting)	5.0 – 8.0	36 – 58
Exhaust pipe to barrel bolts	0.4 – 0.7	3 – 5
Exhaust pipe to frame bolt	1.0 – 1.6	7 – 12
Gearbox drain plug	1.5 – 2.0	11 – 14

1   General description

The crankshaft is a built-up assembly which rotates on two ball journal main bearings and is fitted with needle roller bearings at the connecting rod big- and small-ends. A flywheel generator is fitted on its left-hand end, a gear on its right-hand end providing the drive for the clutch and tachometer. The clutch is a wet multi-plate type mounted on the right-hand end of the gearbox input shaft, the gearbox itself being of the constant mesh type, built in unit with the engine. All engine and gearbox components are housed in the vertically split, aluminium alloy engine unit castings.

The engine/gearbox unit is simple in design and construction requiring the bare minimum of special tools during dismantling and overhaul.

2   Operations with the engine/gearbox unit in the frame

The following items can be removed with the engine/gearbox unit in the frame:

(a)   Carburettor, cylinder head, barrel, piston and reed valve
(b)   Clutch assembly and primary drive pinion
(c)   Kickstart mechanism
(d)   Gearchange shaft mechanism external components
(e)   Flywheel generator assembly
(f)   Gearbox sprocket and neutral switch
(g)   Oil pump

3   Operations requiring engine/gearbox unit removal from the frame

It will be necessary to remove the complete engine/gearbox unit from the frame and to separate the crankcase halves to gain access to the following components:

(a)   Crankshaft, main bearings and oil seals
(b)   Gearbox shafts and bearings, gear selector drum and forks
(c)   Oil pump drive pinion

4   Removing the engine/gearbox unit from the frame

1   If the machine is dirty, it is advisable to wash it thoroughly before starting any major dismantling work. This will make work much easier and also prevent the risk of disturbed lumps of caked-on dirt falling into a vital component during the stripdown.

2   Drain the transmission oil as described in Routine maintenance. While the oil is draining, remove the seat, side panels, frame front covers and fuel tank.

3   Work is made much easier if the machine is lifted to a convenient height on a purpose-built ramp or a platform constructed of planks and concrete blocks. Ensure that the wheels are chocked with wooden blocks so that the machine cannot move and that it is securely tied down so that it cannot fall, also that the stand is supporting it correctly.

4   Disconnect the battery (negative lead first) to prevent any risk of short circuits. If the machine is to be out of service for some time, remove the battery and give it regular refresher charges as described in Chapter 6.

5   Note that whenever any component is removed, refit all mounting nuts, bolts and washers (where fitted) in their original locations for safekeeping.

6   Remove the spark plug cap from the spark plug. Trace the generator lead from the left-hand crankcase to its block connector on the main loom. Disconnect the wires at the connector and the clamp securing them to the frame.

7   Slacken and remove the gearchange lever pinch bolt. Using a centre punch or scriber, mark the position of the gearchange lever as an aid to reassembly then slide the lever off its splines.

8   Remove the four screws securing the left-hand crankcase cover. Remove the cover complete with clutch cable and operating mechanism. It is possible to leave the cover attached to the cable provided it is kept well clear of the engine unit. If separation is required, remove the

4.4 Disconnect the battery terminals to prevent short circuits

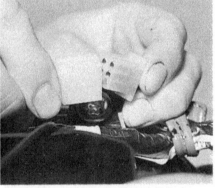

4.6 Disconnect the generator wiring at the block connector

4.7 Remove the gearchange lever having first marked its position on the shaft

4.12 Remove the carburettor rearwards off the barrel studs

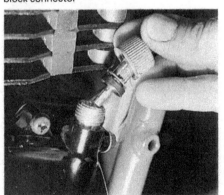

4.13 Release the tachometer cable

4.14a The engine front ...

clutch release inspection cover and disconnect the cable and trunnion from the operating arm.

9   Disconnect the drive chain via its connecting split link and remove the chain. It may be necessary to slacken the chain tension to do this; refer to the chain adjustment section given in Routine maintenance if required.

10   Remove the oil pump cover, slacken the pump cable adjuster and remove the cable. Note the nylon holder at the cable nipple; it is advisable to remove this from the pump lever and tape it to the cable end for safekeeping. Disconnect the oil pump feed pipe from the reservoir tank union and plug its end with a suitably sized clean bolt or screw. Plug the oil tank union with a previously prepared piece of rubber pipe with one end plugged to prevent the loss of oil.

11   Remove the silencer securing bolt and pull the silencer free of the exhaust pipe. Also remove the cylinder barrel mounting and the two frame mountings, then manoeuvre the exhaust system free of the machine. Discard the exhaust port gasket; a new one should be fitted on reassembly.

12   Unscrew the carburettor top, withdraw the throttle slide assembly and secure the throttle/oil pump cable to the frame top tubes, out of harm's way. Pad the throttle slide and needle with clean rag to prevent them from being damaged. Slacken its clamps and remove the air filter hose. Remove the two nuts securing the carburettor to the inlet stub and pull the carburettor rearwards off its mounting studs. Note that the gaskets situated on each side of the insulating spacer should be renewed on reassembly.

13   Unscrew the tachometer cable knurled ring from the right-hand crankcase cover and withdraw the cable.

14   The engine/gearbox unit should now be retained only by its three mounting bolts. Check carefully that all disconnected components will not prevent or hinder the removal of the unit. The engine rear mounting is formed by the swinging arm and its pivot bolt. Remove the retaining nut and washer, and check that the bolt is not seized by turning it with a spanner. The engine is also retained by two other bolts situated at the front and bottom of the engine unit. Remove their retaining nuts and check that they are not seized. If any of the bolts are locked in place by corrosion, apply a liberal quantity of penetrating fluid such as WD40. Allow a short time for it to work then release the bolt by rotating it with a spanner. Working first on the rear mounting, obtain a bolt or length of wooden dowel which is the same diameter as the pivot bolt itself. Use the bolt or dowel as a drift to tap out the pivot bolt until it is clear of the right-hand swinging arm pivot. Pull the pivot bolt out until it just clears the crankcase lug. Do not remove the bolt completely as the swinging arm must be properly supported. Tap out the two remaining bolts using a hammer and soft metal drift.

15   Take care not to damage the alloy engine castings or the painted frame surface when removing the unit. Lift the engine out of the frame and place it on the prepared work surface. Refit the rear mounting bolt so that the swinging arm is fully supported whilst the engine is out of the frame.

4.14b ... and lower mountings

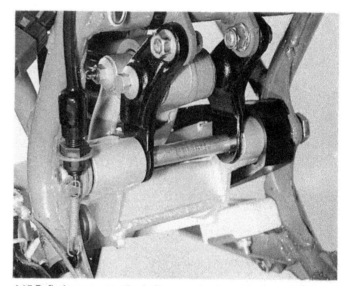

4.15 Refit the rear mounting bolt as soon as the engine is moved so that the rear suspension is supported

5 Dismantling the engine/gearbox unit: preliminaries

1   Before any dismantling work is undertaken, the unit should be thoroughly cleaned. This will prevent the contamination of the engine internals, and will also make working a lot easier and cleaner. A high flash-point solvent, such as paraffin can be used or a proprietary engine degreaser such as Gunk or Jizer. Use old paintbrushes and toothbrushes to work the solvent into the various recesses. Take care to exclude solvent or water from the electrical components and inlet and exhaust ports. The use of petrol should be avoided because of the fire risk.

2   When clean and dry, arrange the unit on the workbench, leaving a suitable clear area for working. Gather a selection of small containers and plastic bags so that parts can be grouped together in an easily identifiable manner. Some paper and a pen should be on hand to permit notes to be made and labels attached where necessary. A supply of clean rag is also required.

3   Before commencing work, read through the appropriate section so that some idea of the necessary procedure can be gained. When removing the various engine components it should be noted that great force is seldom required, unless specified. In many cases, a component's reluctance to be removed is indicative of an incorrect approach or removal method. If in any doubt, re-check with the text.

6 Dismantling the engine/gearbox unit: removing the cylinder head, barrel and piston

1   These components can be removed with the engine/gearbox in or out of the frame. In the former case, the spark plug cap must be disconnected and the carburettor and exhaust system removed.

2   Unscrew and remove the spark plug then, working in a diagonal sequence and by one turn at a time remove the four cylinder head nuts and lift the cylinder head and its gasket off the studs.

3   Using a small screwdriver, lever the oil pipe clamp down the pipe and pull the pipe off its union on the inlet stub. Plug the end of the pipe to prevent oil leakage and the entry of dirt.

4   Remove the two nuts at the rear of the barrel and ease the barrel upwards off the retaining studs. If the barrel appears to be stuck to the crankcase, gently tap around its base with a soft-faced mallet to break the seal of the base gasket. Take care not to damage the fins. If the crankcases are not to be separated, pack the crankcase mouth with a clean rag before lifting the barrel completely off the piston. This will prevent pieces of loose carbon or possibly sections of broken piston ring from falling into the crankcase. Once the barrel is removed, it should be stored upside down to prevent damage to the delicate reed valve assembly.

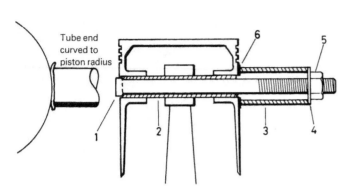

Fig. 1.1 Drawbolt tool used for removing gudgeon pin

1 Drawbolt	4 Washer
2 Gudgeon pin	5 Nut
3 Tubing	6 Rubber washer

5   Use a sharp pointed instrument or a pair of thin nose pliers to remove one of the gudgeon pin retaining circlips. Press out the gudgeon pin far enough to clear the connecting rod and withdraw the piston. Discard the used circlips and obtain new ones for reassembly.

6   If the gudgeon pin is a tight fit in the piston, it will be necessary to warm the piston so that it releases its grip on the gudgeon pin. Taking suitable precautions to prevent scalding, soak a rag in boiling water, wring it out and wrap it around the piston. If the piston is still a tight fit on the gudgeon pin it should be pressed out using a drawbolt tool as shown in the accompanying illustration.

7   Push out the small-end bearing and put it aside for further examination. Remove and discard the barrel base gasket and note the two locating dowels fitted around the studs on the cylinder barrel to crankcase mating surface. These should be removed for safekeeping unless they are pressed tightly into the crankcase castings.

7 Dismantling the engine/gearbox unit: removing the flywheel generator

1   If the engine is in the frame the gear lever and left-hand crankcase cover must first be removed as described in Section 4 and the generator wiring should be disconnected at its block connector.

2   In order to slacken the rotor retaining nut some means must be found of preventing engine rotation. If the engine is in the frame this can be done by locking the engine through the gear train, by selecting top gear and applying the rear brake. If the engine has already been removed from the frame pass a close fitting bar through the connecting rod small-end eye and support it on two wooden blocks placed across the crankcase mouth. Alternatively a holding tool can be fabricated from two lengths of metal which locate in the slots on the flywheel rotor. This tool can be made up using $\frac{1}{8}$ x $1\frac{1}{4}$ in steel strip as shown in the accompanying illustration. Note that the peg bolts should not protrude too far into the rotor slots, otherwise the stator coils may be damaged by contact with the tool. With the engine securely locked, remove the rotor nut.

3   The rotor will be held firmly on the crankshaft taper and will require the use of a centre bolt flywheel puller to draw it safely off the taper; no other method of removal is recommended. A service tool and attachment is available from an authorized Suzuki dealer under part number 09930-30161 and 09930-30102. Alternatively a pattern version can be obtained from most good motorcycle dealers for a reasonable price. Screw the outer body of the puller anticlockwise into the rotor (a left-hand thread is employed) until it seats firmly. Tighten the centre bolt down onto the crankshaft end and tap smartly on the bolt head with a hammer. The shock should jar the rotor free; if not, re-tighten the centre bolt and tap again. Withdraw the rotor and the Woodruff key from the crankshaft end.

4   Disconnect the lead from the neutral indicator switch. Unclip the generator lead from the crankcase and prise out the grommet.

5   Using a scriber or felt marker pen, mark the position of the stator in

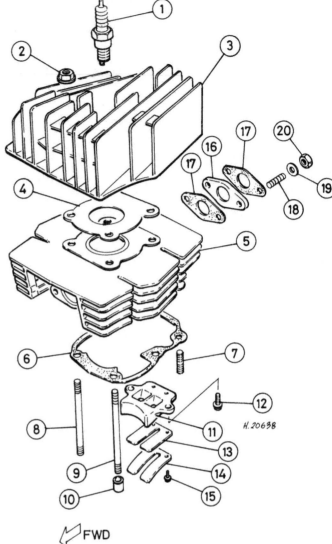

H.20638

FWD

Fig. 1.2 Cylinder head and barrel

1 Spark plug	11 Reed valve block
2 Nut – 4 off	12 Screw – 2 off
3 Cylinder head	13 Reed valve petal
4 Head gasket	14 Stopper plate
5 Cylinder barrel	15 Screw – 2 off
6 Base gasket	16 Insulating spacer
7 Stud – 2 off	17 Gasket – 2 off
8 Stud – 2 off	18 Stud – 2 off
9 Stud – 2 off	19 Washer – 2 off
10 Dowel pin – 2 off	20 Nut – 2 off

relation to the index mark on the crankcase. This will serve as an aid to correct reassembly by preserving the ignition timing position. Note that the manufacturer's timing mark can be highlighted if this corresponds exactly with the casing index mark. Remove the three retaining screws and lift off the stator plate.

8 Dismantling the engine/gearbox unit: removing the neutral indicator switch, gearbox sprocket and oil pump

1   These components can be removed with the engine in or out of the frame. In the former case, the gearchange lever, crankcase left-hand cover and final drive chain must first be removed (see Section 4). Also the oil pump cable should be disconnected and the oil pipes disconnected and plugged to prevent the loss of oil and the entry of air.

6.2 Remove four nuts to release cylinder head

6.4 Cylinder barrel is retained by two nuts at rear

6.5 Prise out gudgeon pin circlips

6.7 If possible remove dowel pins for safekeeping

7.3 Rotor should only be withdrawn using specified tool

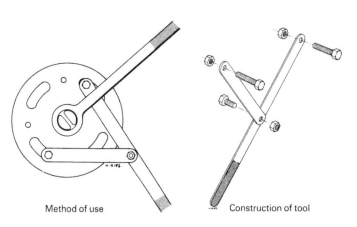

Method of use   Construction of tool

Fig. 1.3 Fabricated flywheel rotor holding tool

7.5 Mark the stator plate position in relation to the casing index mark

2   Remove the two screws securing the neutral indicator switch cover to the crankcase. Lift off the cover and the O-ring beneath it. Withdraw the contact pin and hook out the small spring from its location in the end of the selector drum.

3   The gearbox sprocket is secured to the splined end of the gearbox output shaft by a circlip. Remove the circlip and withdraw the gearbox sprocket from its shaft.

4   Check that the oil pipes have been disconnected and plugged, slacken its two retaining screws and lift off the oil pump. Discard its paper gasket as a new one should be fitted on reassembly.

8.2a Remove its two screws and lift off the neutral switch. Pick out the O-ring ...

8.2b ... contact pin ...

8.2c ... and spring

8.3 Release circlip and withdraw sprocket from output shaft

8.4 Remove two mounting screws and withdraw oil pump

9 Dismantling the engine/gearbox unit: removing the right-hand crankcase cover

1   If the right-hand crankcase cover is to be removed with the engine/gearbox unit installed in the frame the gearbox oil must be drained and the tachometer cable disconnected as described in Section 4 of this Chapter.

2   Remove the kickstart lever pinch bolt and pull the lever off its splined shaft, having previously marked its exact position on the shaft with a centre punch or felt marker pen.

3   Working progressively and in a diagonal sequence, slacken the cover retaining screws and store them in a cardboard template of the cover as a guide to their correct positions on reassembly.

4   Remove the cover, tapping gently with a soft-faced mallet to break the seal and peel off the gasket. Unless the dowel pins are firmly fixed in the crankcase or cover, they should be removed and stored with the cover for safekeeping.

5   If removal or attention to the tachometer drive or driven gear assembly is required, refer to Chapter 4, Section 17.

10.2a Remove bolts, washers and springs ...

10.2b ... and lift off pressure plate

10.4 Once the retaining nut has been removed the outer drum can be withdrawn

10 Dismantling the engine/gearbox unit: removing the clutch assembly and primary drive pinion

1   The clutch assembly and primary drive pinion can be removed with the engine/gearbox unit in or out of the frame. In either case drain the gearbox oil and remove the right-hand crankcase cover as described in the previous Section.

2   Avoid distortion of the clutch pressure plate by slackening its retaining bolts evenly and in a diagonal sequence. Remove the bolts, washers and springs. Withdraw the pressure plate, the thrust washer and headed pushrod. Slide the plain and friction plates off the centre as a set.

3   The clutch centre must be locked to permit the retaining nut to be slackened. If the engine is in the frame, select top gear and apply the rear brake, locking the transmission. Alternatively, make up a holding tool from two strips of metal and a nut and bolt as shown in the accompanying illustration and photo 33.5b. With the clutch centre locked, flatten the tab washer and remove the retaining nut.

4   Lift off the clutch centre and the thrust washer, followed by the clutch outer drum, bush and second thrust washer. Withdraw the long pushrod from the input shaft. Store all the clutch components together whilst noting their exact relative positions to assist reassembly.

5   Lock the crankshaft by passing a close fitting bar through the small-end eye and resting it on two wooden blocks placed across the crankcase mouth. Remove the primary drive pinion retaining nut, washer, the primary drive pinion, Woodruff key and the spacer, noting their exact relative positions to aid refitting.

11 Dismantling the engine/gearbox unit: removing the kickstart assembly and idler gear

1   The kickstart shaft can be removed with the engine in or out of the frame. It will first be necessary to drain the transmission oil and remove the crankcase right-hand cover as described in Section 9. Note that if removal of the idler gear is required the clutch must be removed for access.

2   Disengage the hooked end of the kickstart return spring from the spring post set in the crankcase and carefully allow it to unwind until the tension is released.

3   Withdraw the plastic spring guide, unhook the spring and remove it from the kickstart shaft. Remove the spring support washer and lift off the kickstart drive gear complete with its spring clip. The shaft can be pulled out of its bore in the crankcase.

4   Removal of the clutch assembly is necessary for access to the kickstart idler gear, situated on the end of the output shaft. Release its retaining circlip and withdraw the thrust washer, idler gear and wave washer from the output shaft.

12 Dismantling the engine/gearbox unit: removing the gearchange shaft and selector ratchet

1   The gearchange shaft and selector ratchet can be removed with the engine/gearbox unit in or out of the frame. In the former case the

10.5 Remove nut and withdraw primary drive pinion, key and spacer

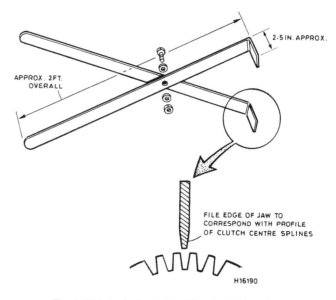

Fig. 1.4 Fabricated clutch holding tool dimensions

12.2 Pull gearchange shaft out of crankcase

12.3 Remove pawl guide plate and ratchet locating plate

13.3 Remove crankcase retaining screws to separate crankcase halves

gearchange pedal, right-hand crankcase cover and clutch must be removed as described in earlier Sections of this Chapter. Note that attention to the selector drum, forks and dentent mechanism will necessitate separation of the crankcases as described in the following section.

2   Pull the gearchange shaft through the crankcase from the right-hand side. Take care when the splines on the shaft end pass through the oil seal in the left-hand crankcase; it is good practice to wrap some tape over the shaft splines to protect the oil seal.

3   Withdraw the pawl guide plate and ratchet locating plate by removing their countersunk retaining screws. The use of an impact driver may be necessary as these screws are thread locked into position.

4   The ratchet pawls are spring loaded and will fly out if not removed carefully. Squeeze the pawls together gently against spring tension and lift the complete assembly out of the selector drum. Wrap an elastic band around the ratchet to prevent the loss of the springs, plungers and pawls.

13   Dismantling the engine/gearbox unit: separating the crankcase halves

1   The crankcases can only be separated once the engine/gearbox unit has been removed from the frame and all preliminary dismantling operations described in Section 5 – 12 of this Chapter have been carried out.

2   Support the crankcase on wooden blocks so that the crankshaft end and exposed gearbox shafts are clear of the working surface. Make a final check that all components have been removed which might hinder crankcase separation.

3   Remove the nine crankcase fastening screws, slackening them progressively and in a diagonal sequence from the outside inwards. It is a good idea to make up a cardboard template through which the screws can be inserted as they are removed, thus preventing their loss and ensuring that they are refitted in their correct locations.

4   Holding the two halves firmly together, invert the crankcase and resupport it on the wooden blocks. This will ensure that the crankshaft and gearbox components remain in place in the left-hand crankcase half after the right-hand half is removed.

5   Using only a soft-faced mallet, tap gently on the exposed ends of the crankshaft and gearbox shafts and all around the joint area of the two crankcase halves until initial separation is achieved. Lift off the right-hand half, keeping it absolutely square so that the bearings do not stick on their respective shafts; tap gently on the shaft ends to assist removal. **Do not** use excessive force and **never** attempt to lever the cases apart. If difficulty is encountered, tap the cases back together and start again.

6   Difficulty in achieving initial separation may be due to corrosion on the locating dowel pins. Apply a liberal coating of penetrating fluid to the joint area and inside the passages through which the crankcase screws are fitted. Allow time for it to work and start again.

7   Suzuki provide a special tool, Part number 09920-13120, which can be used to separate the crankcase halves should all other attempts fail.

8   When the crankcases have been separated, ensure that all internal components remain in their correct positions in the left-hand casing.

9   If they are not pressed tightly into their bores, remove the two locating dowels and store them safely with the right-hand crankcase half.

14   Dismantling the engine/gearbox unit: removing the crankshaft and gearbox components

1   Remove the stopper arm spring, pull out both selector fork shafts and manoeuvre the selector forks out of their tracks in the selector drum. Withdraw the drum and camplate taking care not to lose the small pin located between the two components. Remove the selector forks and stopper arm, using a felt marker to mark each fork with its position in relation to the gear shafts and selector drum. It is advisable to refit the selector forks on their respective shafts as a guide to reassembly.

2   Lift off the white plastic oil pump drive pinion. Remove both shafts and gear clusters as a single unit. If either shaft appears to be stuck in its bearing, a gentle tap to the shaft end with a soft-faced mallet should be sufficient to jar it free.

3   Invert the crankcase half and use wooden blocks to keep the

crankshaft end clear of the working surface. Position the blocks as close as possible to the crankshaft to provide maximum support for the crankcase casting and ensure that they are tall enough to allow the crankshaft to be removed from under the casting. Protect the end of the shaft by refitting the rotor retaining nut until it is flush with the end of the shaft. Use a soft-faced mallet to drive the crankshaft down and release its grip on the bearing. Be careful not to allow the crankshaft to drop as soon as it becomes free. Remember to remove the nut before completely separating the crankshaft from the crankcase.

4   If difficulty is encountered when attempting to remove the crankshaft do not use excessive force in order to displace it. Take the assembly to an authorized Suzuki dealer for the crankshaft to be removed using Suzuki's own service tool.

5   To allow examination and, where required, removal of the crankshaft and input shaft bearings, the retaining cover plates should first be removed from the right-hand crankcase half. As the screws are thread-locked, the use of an impact driver will probably be required.

15   Dismantling the engine/gearbox unit: removing oil seals and bearings

1   Oil seals are easily damaged when disturbed and should be renewed as a matter of course during overhaul. Lever them out of position

15.1 Oil seals can be levered out of castings

15.2 Castings should be heated to permit removal of bearings

using a flat-bladed screwdriver and take care not to damage the alloy seal housing; note which way round the seals are fitted.

2 The crankshaft and gearbox bearings are a press fit in their respective locations. Remove them by placing the crankcase in an oven and heating it to approximately 100°C or by immersing it in boiling water. The casing may then be sharply tapped, face down, on a wooden surface to jar the bearings free. Alternatively the bearings may be drifted out of position using a suitably sized drift. Apply the drift only to the bearing outer race and ensure that it leaves its housing squarely.

3 Always take care to prevent personal injury when handling heated components.

16 Examination and renovation: general

1 Before any component is examined, it must be cleaned thoroughly. Being careful not to mark or damaged the item in question, use a blunt-edged scraper (an old kitchen knife or a broken plastic ruler can be very useful) to remove any caked-on deposits of dirt or oil, followed by a good scrub with a soft wire brush (a brass wire brush of the type sold for cleaning suede shoes is best, with an assortment of bottle-cleaning brushes for ports, passages etc). Take care not to remove any paint code marks from internal components.

2 Soak the component in a solvent to remove the bulk of the remaining dirt or oil. If one of the proprietary engine degreasers (such as Gunk or Jizer) is not available, a high flash-point solvent such as paraffin (kerosene) should be used. The use of petrol as a cleaning agent cannot be recommended because of the fire risk. With all of the above cleaning agents take great care to prevent any drops getting into the eyes and try to avoid prolonged skin contact. To finish off the cleaning procedure wash each component in hot soapy water (as hot as your hands can bear); this will remove a surprising amount of dirt on its own and the residual heat usually dries the component very effectively. Carefully scrape away any remaining traces of old gasket material from all joint faces.

3 Check all ports, passages and oilways for blockages, using compressed air to clear them, or implements such as pipe cleaners.

4 If there is the slightest doubt about the lubrication system, for example if a fault appears to have been caused by a failure of the oil supply, all components should be dismantled so that the oilways can be checked and cleared of any possible obstructions. Always use clean, lint-free rag for cleaning and drying components to prevent the risk of small particles obstructing oilways.

5 Examine each part carefully to determine the extent of wear, checking with the tolerance figures listed in the Specifications section of this Chapter. If there is any doubt about the condition of a particular component, play safe and renew.

6 Various instruments for measuring wear are required, including an internal and external micrometer or vernier gauge, and a set of standard feeler gauges. Additionally, although not absolutely necessary, a dial gauge and mounting bracket are invaluable for accurate measurement of endfloat, and play between components of very low diameter bores – where a micrometer cannot reach. After some experience has been gained, the state of wear of many components can be determined visually, or by feel, and a decision on their suitability for re-use can be made without resorting to direct measurement.

17 Examination and renovation: crankcases and covers

1 Careful examination should be made of the aluminium alloy castings for signs of damage. Any cracks, broken fragments or holes will require professional treatment or renewal, although some may be temporarily repaired with an epoxy resin adhesive such as Araldite. Damage around the bearing bosses will normally indicate that renewal is necessary because a small discrepancy in these areas can result in serious mis-alignment of the shaft concerned.

2 Should any internal threads require repair, now is the time to deal with them. Tapping out threads to a convenient size or where this is not possible the use of a diamond-section wire insert of the Helicoil type are the most common methods of repair.

3 Sheared bolts or screws can usually be removed with a screw extractor which consists of a left-hand thread tapered high tensile steel screw. When inserted anticlockwise into a pre-drilled hole, the sheared

bolt or screw is withdrawn.

4 If a problem arises which seems beyond your scope, it is well worth consulting a professional such as a local engineering company or motorcycle dealer before condemning an otherwise sound casting.

5 If gasket or other mating surfaces are marked or damaged in any way they can be reclaimed by rubbing them on a sheet of fine abrasive paper laid on an absolutely flat surface such as a sheet of plate glass. Use a gentle figure-of-eight pattern, maintaining light but even pressure on the casting. Note that if large amounts of material are to be removed, advice should be sought as to the viability of re-using the casting in question; the internal clearances are minimal in many cases between rotating or moving components and the castings. Stop work as soon as the entire mating surface is polished by the action of the paper.

6 Large surfaces such as the crankcase surface will have to be skimmed on a surface plate if warped. This is a task for a light engineering business only; be careful to warn them to remove only the minimum amount of metal necessary to true up the face. If excessive warpage is found, seek expert advice.

7 Note that the mating surface may become distorted outwards around the mounting screw holes, usually because these have been grossly overtightened. In this event, use a large drill bit or countersink to very lightly skim the raised lip from around the screw hole, then clean up the whole surface as described above.

8 Finally, check that all screw or bolt tapped holes are clean down to the bottom of each hole; serious damage can be caused by forcing a screw or bolt down a dirty thread and against an incorrect stop caused by the presence of dirt, oil, swarf or blobs of old jointing compound. At the very least the component concerned will be incorrectly fastened, at worst the casting could be cracked. The simplest way of cleaning such holes is to use a length of welding rod or similar to check that the hole is clean all the way to the bottom and to dig out any embedded foreign matter, then to give each hole a squirt of contact cleaner or similar solvent applied from an aerosol via the long plastic nozzle usually supplied. Be careful to wear suitable eye protection while doing this; the amount of dirt and debris that can be ejected from each hole is surprising.

18 Examination and renovation: bearings and oil seals

1 The crankshaft and gearbox bearings can be examined while they are still in place in the crankcase castings. Wash them thoroughly to remove all traces of oil, then feel for free play by attempting to move the inner race up and down, then from side- to-side. Examine the bearing walls or rollers and the bearing tracks for pitting or other signs of wear, then spin the bearing hard. Any roughness caused by the defects in the bearing balls or rollers or in the bearing tracks will be felt and heard immediately.

2 If any signs of free play are discovered, or if the bearing is not free and smooth in rotation but runs roughly and slows down jerkily, it must be renewed. Bearing removal is described in Section 15 of this Chapter, and refitting in Section 28.

3 To prevent oil leaks occurring in the future, all oil seals and O-rings should be renewed whenever they are disturbed during the course of an overhaul, regardless of their apparent condition. This is particularly true of the main bearing oil seals, which are a weak point on any two-stroke.

19 Examination and renovation: crankshaft assembly

1 Big-end failure is characterised by a pronounced knock which will be most noticeable when the engine is worked hard. Crankshaft wear is only likely to take place after a very high mileage has been covered or from failure of the lubrication supply. The general condition of the crankshaft big-end bearing can be established with the assembly removed from the engine, or with the cylinder head and barrel removed.

2 Check for wear with the crankshaft set in the TDC (top dead centre) position by pulling and pushing the connecting rod. No discernible movement will be evident in an unworn bearing but care must be taken not to confuse endfloat, which is normal, and bearing wear. If a dial gauge is available, further tests can be carried out by setting the gauge pointer so that it abuts against the upper edge of the periphery of the small-end eye. Measurement can then be taken of the amount of side to side deflection of the connecting rod. If this measurement exceeds the

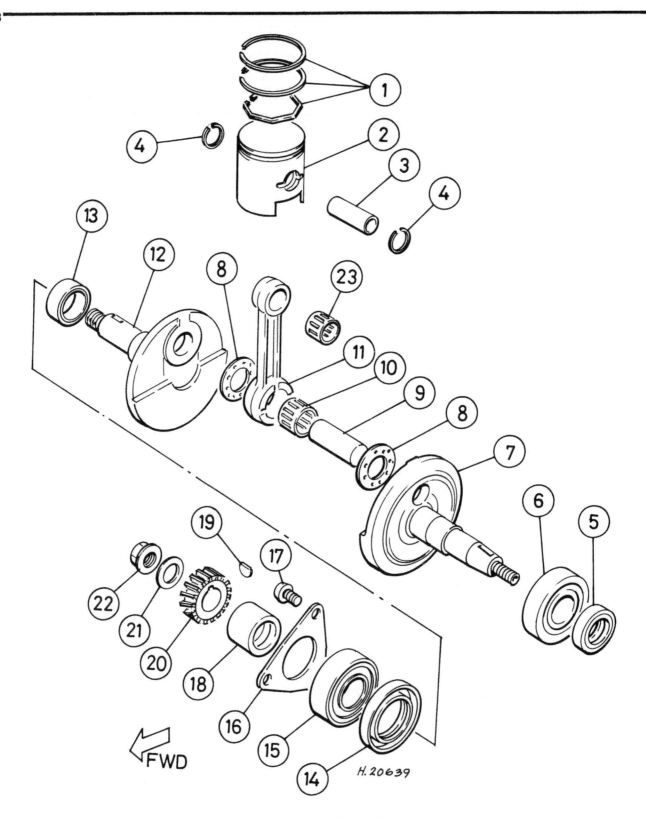

Fig. 1.5 Crankshaft assembly

1	Piston rings	7	Left-hand flywheel	13	Spacer	19 Woodruff key
2	Piston	8	Thrust bearing – 2 off	14	Right-hand oil seal	20 Primary drive pinion
3	Gudgeon pin	9	Crank pin	15	Right-hand main bearing	21 Lock washer
4	Circlip – 2 off	10	Big-end bearing	16	Bearing retainer plate	22 Nut
5	Left-hand oil seal	11	Connecting rod	17	Screw – 3 off	23 Small-end bearing
6	Left-hand main bearing	12	Right-hand flywheel	18	Spacer	

H.20639

FWD

service limit, the big-end bearing must be renewed.

3   Measurement of crankshaft runout can only be made with the crankshaft assembly removed from the crankcase and set up on V-blocks which themselves have been positioned on a completely flat surface. The amount of runout should be measured with a dial gauge at a point just inboard of the threaded portion of the left-hand mainshaft. If the amount of runout shown by the gauge needle exceeds the service limit the assembly must be renewed.

4   Like the big-end bearing, the small-end bearing is of the caged needle roller type and will seldom give trouble unless a lubrication failure has occurred. Push the bearing into the small-end eye of the connecting rod and push the gudgeon pin through the bearing. Hold the connecting rod steady and feel for any discernible movement between it and the gudgeon pin. If movement is felt, do not automatically assume that the bearing is worn but check that the bore of the small-end eye and the outer diameter of the gudgeon pin are not worn beyond their service limits (given in the Specifications Section). If both the connecting rod and gudgeon pin are found to be within limits, discard the bearing and replace it with a new item. Close inspection of the bearing will show if the roller cage is beginning to crack or wear, in which case the bearing must be renewed.

5   Make a careful check, using a magnifying glass if necessary, for any obvious signs of wear or damage such as cracks or bending in the connecting rod, scratched or scored mainshafts (especially at those points against which the oil seal lips bear) and damage to the threads on the mainshafts ends and damage to the keyways. If any fault is found or suspected in any of the components comprising the crankshaft assembly, it is recommended that the complete crankshaft assembly is taken to an authorized Suzuki dealer, who will be able to confirm the worst, and supply a new or service-exchange assembly. The task of dismantling and reconditioning the big-end assembly is a specialist task, and it is considered to be beyond the scope and facilities of the average owner.

---

20   Examination and renovation: cylinder barrel

1   The usual indication of a badly worn cylinder barrel and piston is piston slap, a metallic rattle that occurs when there is little or no load on the engine.

2   Wash the cooling fins on the cylinder barrel thoroughly so that they are free from traces of mud and dirt, as clogged fins can cause the engine to overheat. Remove any build-up of carbon from the exhaust port with a small scraper and finish off with a metal polish to give a smooth finish. Ensure that any traces of old gasket have been removed from the cylinder base and head mating surfaces and carburettor mounting.

3   Make a close visual examination of the bore surface, looking for signs of scoring or other damage, particularly if piston or piston ring damage has also been discovered. Any damage of this nature will necessitate reboring and a new piston, as it is impossible to obtain a good seal if the bore is not perfectly finished.

4   Carefully remove the ring of carbon from the bore mouth so that the amount of bore wear can be correctly assessed. There will probably be a lip, at the mouth which marks the limit of travel of the top piston ring. The depth of this lip will give a good indication of the amount of bore wear even though the amount of wear is not evenly distributed.

5   The most accurate method of measuring bore wear is by the use of a cylinder bore DTI (Dial Test Indicator) or a bore micrometer. Measurement should be taken at a point 15 mm (0.59 in) from the top of the cylinder bore and the reading obtained compared with the service limit for bore wear given in the Specifications Section. Note that it is necessary to rotate the measuring instrument so that the point of greatest wear in the bore is found.

6   In the absence of a bore micrometer, a rough indication can be made by inserting the piston, less rings, into the bore. Use feeler gauges to measure the gap between the piston and bore at several points with the piston at the top of the bore and then moved down to the bottom of the bore. Then subtract the smallest feeler gauge thickness from the largest to obtain the wear figure. Note that this method is far from accurate, since it does not allow for the curved surfaces of the two components being measured. It is recommended that the measurement is checked to confirm the need for a rebore.

7   If it is found that the amount of wear in the bore exceeds those limits

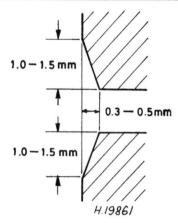

Fig. 1.6 Cylinder bore port chamfer dimensions

given, then it will be necessary to have the cylinder barrel rebored to the next oversize and the appropriate oversize piston fitted.

8   On receiving the cylinder barrel back from the reborer, check the edges of the ports at the bore end to ensure that they have been correctly chamfered to the measurements given in the accompanying illustration. This work must be done before the cylinder barrel is refitted to the machine, otherwise there is a distinct possibility that the piston rings will catch on the unchamfered edge of each port and break, thus necessitating a further strip down and rebore. Chamfering of the port edges can be carried out by very careful use of a scraper but it is essential to ensure that the wall of the bore does not become damaged in the process. Finish off the process by polishing the cut edges with fine emery paper.

9   Checking the piston/cylinder clearance can be made either by direct measurement of the cylinder bore and piston diameter and then by subtracting the gap, or by using a feeler gauge. In either case, if the clearance exceeds the maximum service limit there is evidence that a new piston or a rebore and new piston is required. Note that if the method of direct measurement of the piston and bore is decided upon, then the measurement for piston diameter should be made at a point 23 mm (0.90 in) from the base of the piston skirt, at right-angles to the gudgeon pin hole, whereas the measurement for cylinder bore diameter should be made at various points around a line 15 mm (0.59 in) from the top of the bore.

10   Check the cylinder barrel to cylinder head mating surface for distortion by placing a straightedge across several places on it and attempting to slide a 0.05 mm (0.0020 in) feeler gauge between the straightedge and mating surface. If the cylinder barrel proves to be warped beyond this limit, grind it flat by placing a sheet of emery paper on a surface plate or sheet of plate glass and rubbing the mating surface against it, in a slow, circular motion. Commence this operation with 200 grade paper and finish with 400 grade paper and oil. If it is thought necessary to remove a substantial amount of metal in order to bring the mating surface back to within limits, obtain advice from an authorized Suzuki dealer as to whether it is necessary to renew the barrel.

11   Finally, take great care not to damage the delicate reed valve assembly when handling the cylinder barrel. If it is to be removed, ensure that it is cleaned and free from damage and that its retaining screws are thread-locked into position upon refitting. See Chapter 2, Section 9.

12   If the original bore is to be re-used along with a new piston or piston rings, the surface should be honed to remove the shiny glazed surface. More importantly, the wear ridge at the top of the bore must be removed or the new top ring may break when it strikes it. This is best entrusted to a professional, who will have the necessary de-ridging and honing facilities.

---

21   Examination and renovation: cylinder head

1   Check that the cylinder head fins are not clogged with oil or road dirt, using an old toothbrush and degreasing agent if necessary. Examine the cylinder head for signs of cracking, especially around the spark plug and stud holes.

2   Check the condition of the thread in the spark plug hole. If it is
damaged an effective repair can be carried out using a diamond section
wire thread insert of the Helicoil type. Any deposits of carbon should be
removed from the combustion area of the cylinder head as described
under the decarbonisation section in Routine maintenance.

3   If leakage has occurred between the mating surfaces of the cylinder
head and barrel, check for distortion as described in Section 20 of this
Chapter and take the appropriate remedial course of action. Note that
most cases of distortion are due to overtightening or uneven tightening
of the cylinder head nuts; always adhere to the specified torque setting.

---

22   Examination and renovation: piston and piston rings

1   If a rebore is necessary, the existing piston and rings can be
disregarded because they will be replaced with their oversize equiva-
lents as a matter of course.

2   Remove the piston rings by pushing the ends apart with the thumbs
whilst gently easing each ring from its groove. Great care is necessary
throughout this operation because the rings are brittle and will break
easily if overstressed. If the rings are gummed in their grooves, three
strips of tin can be used, to ease them free, as shown in the accompanying illustration. Take great care to keep the rings separate and the right
way up so that they can be refitted in their correct positions. Note the
expander ring fitted behind the second ring.

3   Piston wear usually occurs at the skirt or lower end of the piston and
takes the form of vertical stroke or score marks on the thrust side. There
may also be some variation in the thickness of the skirt. Measurement
for piston diameter should be taken at a point 23 mm (0.90 in) from the
base of the piston skirt, at right-angles to the gudgeon pin hole. If the
measurement obtained is found to be less than the service limit given in
the Specifications Section, then the piston must be renewed.

4   Check that the piston and bore are not scored, particularly if the
engine has tightened up or seized. If the bore is badly scored, it will
require a rebore and oversize piston. If the scoring is not too severe or
the piston has just picked up, it is possible to remove the piston high
spots by careful use of a fine swiss file. Application of chalk to the file

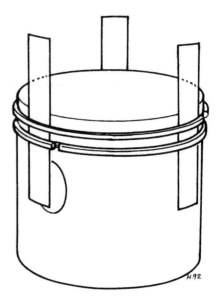

Fig. 1.7 Method of removing gummed piston rings

teeth will help prevent clogging of the teeth, and the subsequent risk of
scoring.

5   Check for any build-up of carbon in the piston ring grooves. Any
carbon should be carefully removed by using a section of broken piston
ring or similar. The piston ring grooves may have become enlarged in
use, thus allowing each ring to have a greater clearance than is permissible. To measure the ring to groove clearance insert each piston ring in its
cleaned groove and measure the clearance between the side of the ring
and the groove with a feeler gauge. If the measurement obtained
exceeds the specified clearance the piston is due for renewal. Note that
Keystone-type piston rings are employed. For this reason the feeler
gauge must be inserted between the lower, flat surface of the piston

22.3 Measure piston diameter at right-angles
to gudgeon pin

22.5 Measuring piston ring-to-groove
clearance

22.9a Ensure ring is positioned squarely in
bore when measuring installed end gap

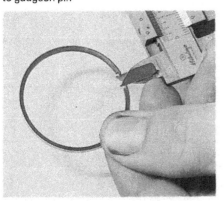

22.9b Measuring piston ring free end gap

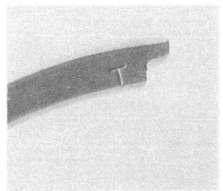

22.9c Ring top face will be marked with either
R or T

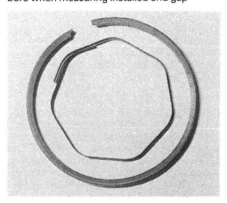

22.10 Note expander ring fitted to second ring
groove

ring and the bottom edge of the ring groove to obtain an accurate measurement.

6 When cleaning the piston ring grooves, check that the ring locating peg located in each groove is not loose or worn. If in doubt as to the condition of these pegs, seek professional advice from an authorized Suzuki dealer and renew the piston if necessary.

7 The gudgeon pin should be a firm press fit in the piston. Check for scoring on the bearing surfaces of each part and where damage or wear is found, renew the part affected. Note that the degree of wear can be assessed by direct measurement if the equipment is available. The pin circlip retaining grooves must be undamaged; renew the piston rather than risk damage to the bore through a circlip becoming detached. Discard the circlips themselves; these should **never** be re-used.

8 Examine the working surface of each piston ring. If discoloured areas are evident, the ring should be renewed because these areas indicate the blow-by of gas.

9 Piston ring wear is measured by inserting each ring in a part of the cylinder bore which is not normally subject to wear. Ideally, the ring should be inserted into the cylinder bore so that it is positioned near the bottom of the bore. Use the crown of the piston as a means of locating the ring squarely in the bore and measure the gap between the ring ends with a feeler gauge. If the gap measured is found to be greater than the service limit, then the ring must be renewed. Note that ring end gap service limits are given also for when each ring is in its free state. These are given as a means of determining the spring tension of each ring. Measure this end gap with the ring placed flat on a clean work surface and compare the reading obtained with the appropriate reading given in the Specifications Section. The piston rings are marked on their upper surface by the letter 'R' or 'T' depending on the manufacturer.

10 Bear in mind, when refitting used rings to the piston, that they must be fitted in exactly the same positions as noted during removal. When fitting either new or old rings, take note of the following points. Each ring should be fitted to the piston by pulling its ends apart just enough to allow it to pass over the piston crown and into its groove. Always fit each top and second ring with its marked (R or T) surface uppermost and remember to fit the expander ring behind the second ring. Check that each ring presses easily into its groove and that its ends locate correctly over the locating pin.

11 Finally, do not automatically assume when fitting new rings, that their end gaps will be correct. As with part worn rings, the end gap must be measured. It may be necessary to enlarge the gap, in which case this should be done by careful use of a needle file. With the rings fitted to the piston, place the assembly to one side ready for engine reassembly.

12 It must be noted that when fitting new piston rings to a used piston which is to be run in a part-worn cylinder bore, the bore surface must be prepared first by glaze-busting as described in Section 20.

23 Examination and renovation: gearbox components

1 Give each of the gearbox components a close visual inspection for signs of wear or damage such as broken or chipped teeth, worn dogs, damaged or worn splines or bent selectors. Renew any damaged or worn components; repairs are impracticable.

Gear shaft assemblies

2 If dismantling of the input shaft assembly proves to be necessary, the 2nd gear will have to be pressed from position by using a hydraulic press; no other method of removal is possible. As it is unlikely that this type of tool will be readily available, it is recommended that the complete shaft assembly be returned to an authorized Suzuki dealer who will be able to remove the gear, renew any worn or damaged components and return the shaft assembly complete.

3 If an hydraulic press is available and it is decided to attempt removal of the 2nd gear from the input shaft, it is very important to fully realise the dangers involved when using such a tool. Both the tool and the shaft assembly must be set up so that there is no danger of either item slipping. The tool must be correctly assembled in accordance with the maker's instructions as the force exerted by the tool is considerable and perfectly capable of stripping any threads from holding studs or inflicting other damage upon itself and the input shaft. Always wear proper eye protection in case a component should fail and shatter before it becomes free, as will happen if the component is flawed.

4 With the 2nd gear withdrawn, slide the combined 3rd/4th gear off the shaft end. Release its retaining circlip to permit removal of the 5th gear. The 1st gear is integral with the input shaft and if damaged will necessitate the complete renewal of the shaft.

5 The output shaft can be easily dismantled without the aid of a press. Slide the thrust washer and 1st gear off its right-hand end. Release its retaining circlip and remove the 5th gear. Remove the circlip, splined thrust washer, 4th and 3rd gears, followed by a further splined thrust washer. The sliding dog and 2nd gear are both independently retained by circlips.

6 When removing the shaft circlips only open them sufficiently to be slid over the shaft and renew any which are distorted or a loose fit on their shafts. Note that the manufacturer recommends that the shaft circlips be renewed as a matter of course whenever the shafts are dismantled. Also inspect the thrust washers and renew any that are damaged.

7 Note that certain gears have a bush fitted within their centres. If any one of these bushes appears to be worn or in any way damaged, then the gear should be returned to an authorized Suzuki dealer who will be able to advise on which course of action to take as to its renewal.

8 Carefully inspect the splines on both shafts and gears for any signs of wear, hairline cracks or breaking down of the hardened surface finish. If any one of these defects is apparent, then the offending component must be renewed. It should be noted that damage and wear rarely occur in a gearbox which has been properly used and correctly lubricated, unless very high mileages have been covered.

9 It should be assumed that the gearbox sprocket constitutes part of the output shaft assembly and should therefore be examined along with the rest of that assembly. Clean the sprocket thoroughly and examine it closely, paying particular attention to the condition of the teeth. The sprocket should be renewed if the teeth are hooked, chipped, broken or badly worn. It is considered bad practice to renew one sprocket on its own; both drive sprockets should be renewed as a pair, preferably with a new final drive chain. If this recommendation is not observed, rapid wear resulting from the running of old and new parts together will necessitate even earlier renewal on the next occasion. Examine the splined centre of the sprocket for signs of wear. If any wear is found, renew the sprocket as slight wear between the sprocket and shaft will rapidly increase due to the torsional forces involved. Remember that the output shaft will probably wear in unison with the sprocket, it is therefore necessary to carry out a close inspection of the shaft splines.

10 Rebuild the gear shafts as described in Section 27, noting that the input shaft 2nd gear must be pressed onto the shaft to a prescribed depth.

Selector drum and gearchange mechanism

11 Examine the selector forks carefully, ensuring that there is no sign of scoring on the bearing surface of either their fork ends, their bores or their gearchange drum locating pins. Check for any signs of cracking

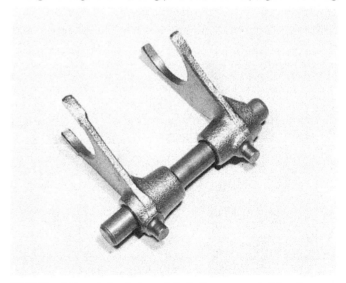

23.11 Check the selector forks and shafts for wear on their bearing surfaces

around the edges of the bores or at the base of the fork arms. Refer to the Specifications Section and measure the thickness of the fork claw ends; renew the fork if the measurement obtained is less than the limit given.

12   Place each selector fork in its respective pinion groove and using a feeler gauge, measure the claw end to pinion groove clearance. If the measurement obtained exceeds the service limit, then it must be decided whether it is necessary to renew one or both components. The acceptable limits for pinion groove width are given in the Specifications Section of this Chapter.

13   Check each selector fork shaft for straightness by rolling it on a sheet of plate glass and checking for any clearance between the shaft and the glass with feeler gauges. A bent shaft will cause difficulty in selecting gears. There should be no sign of any scoring on the bearing surface of the shaft or any discernible play between each shaft and its selector fork(s).

14   The tracks in the selector drum, which co-ordinate the movement of the selector forks, should not show signs of undue wear or damage.

15   Examine the camplate, stopper arm and its return spring for wear, paying particular attention to the roller on the stopper arm end which should be free from cracks and flat spots. If the selector rachet and pawl assembly show signs of burring or wear, the damaged items must be renewed as wear on the pawl ends can give rise to many gear selection problems. The gearchange shaft is only likely to cause difficulties if it is bent, usually as a result of the machine being dropped onto its gear lever. Check that the splines on its end are not damaged and that the return spring is in a serviceable condition. If any doubt arises about the condition of a gear selector component, comparison with a new part will show the amount of wear that has taken place.

Oil pump drive pinion
16   Look closely at the oil pump drive pinion for cracks in the plastic or broken, chipped or worn teeth. Note that failure of this part will lead to complete engine seizure and costly repairs.

---

24   Examination and renovation: clutch and primary drive

---

1   Clean the clutch components with a high flash-point solvent to remove all traces of friction material and sludge.

2   Measure the thickness of each friction plate and renew as a complete set if the friction material has worn to the service limit given in the Specifications Section. At the same time examine the tongues on each friction plate and the corresponding slots in the clutch outer drum for indentations caused by clutch chatter. If slight, the damage can be removed with a fine file but ensure that the tongue width does not exceed its service limit.

3   Look for distortion in each plain plate by laying it on a flat surface such as a sheet of plate glass whilst attempting to insert a 0.1 mm (0.0039 in) feeler gauge between the plate and the flat surface. Each plain plate must be free from signs of scoring and overheating in the form of blueing. Check that the plate thickness is within the tolerance given in the Specifications and renew accordingly. As with the friction plates, check the tongues on the inner periphery of the plain plates against their channels in the clutch centre and remove any slight damage with a fine file.

4   Measure the free length of each clutch spring. Renew as a set if any

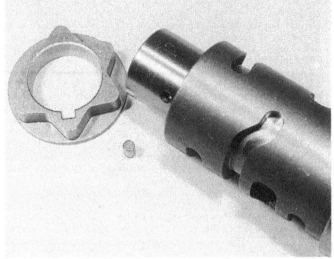

23.14 Examine the camplate and selector drum tracks for wear

one has set to the service limit given in the Specifications.

5   Check that the thrust bearing is in good condition and renew it if there is any damage to the rollers or cage. Also examine the short headed pushrod and the long pushrod, looking for wear on all points of contact. The clutch release mechanism is situated in the crankcase left-hand cover, access for adjustment being via a separate inspection cover. Apart from ensuring that the operating worm is greased, no attention should be necessary to the mechanism; if required, it can be detached from the crankcase cover by releasing the two retaining screws.

6   Examine the teeth of the primary drive pinion and the corresponding teeth of the large gear mounted on the back of the clutch outer drum. If there is damage to either pinion, they should be renewed as a pair due to both sets of teeth being worn in unison.

7   Wear or damage to the Woodruff key on the crankshaft right- hand end is rare but if apparent, it will invariably mean that the keyways in the pinion and crankshaft will be similarly affected. Apart from renewing these components, the only satisfactory answer to the problem of worn or damaged keyways is either to have them recut to an acceptably larger size and have a new key fitted or to have each keyway refilled by either a welding or metal spraying process and then recut to accommodate a key of the original size. Either one of these alternatives requires a high degree of skill in the use of specialist equipment and the work should, therefore, be placed in the hands of a light engineering company which specialises in such a task.

---

25   Examination and renovation: kickstart assembly

---

1   Wear on the kickstart components is only likely after a considerable

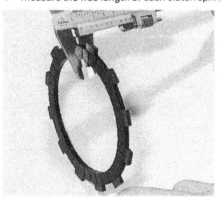

24.2a Measuring the clutch friction plate thickness ...

24.2b ... and the tongue width

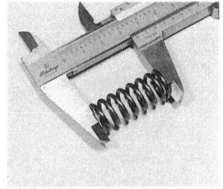

24.4 Measuring the clutch spring free length

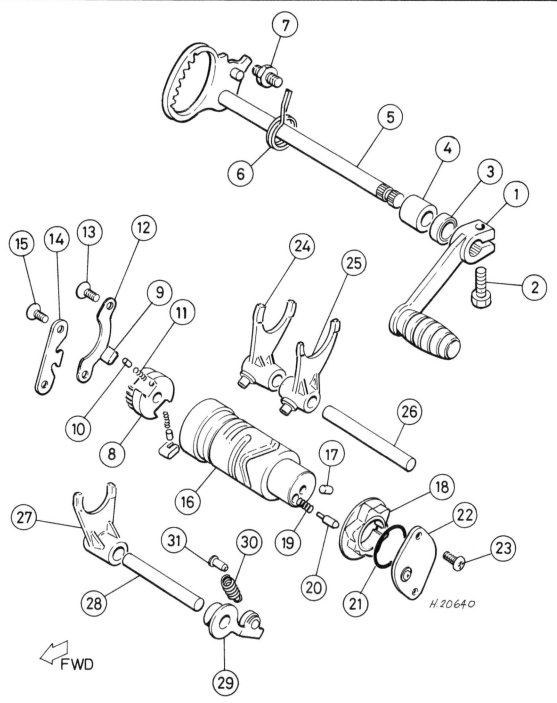

Fig. 1.8 Selector drum and gearchange mechanism

1 Gear lever	9 Pawl – 2 off	17 Camplate locating pin	25 Output shaft sliding dog fork
2 Pinch bolt	10 Plunger – 2 off	18 Camplate	26 Fork shaft
3 Oil seal	11 Spring – 2 off	19 Spring	27 Input shaft fork
4 Spring guide	12 Pawl guide plate	20 Neutral contact pin	28 Fork shaft
5 Gearchange shaft	13 Screw – 2 off	21 O-ring	29 Stopper arm
6 Return spring	14 Ratchet guide plate	22 Neutral switch	30 Return spring
7 Spring post	15 Screw – 2 off	23 Screw – 2 off	31 Spring post
8 Ratchet	16 Selector drum	24 Output shaft 5th gear fork	

mileage has been covered, with a fatigued or damaged kickstart return spring being the most likely component to fail.

2   Carefully clean and inspect each item, looking for worn or chipped gear teeth, damaged splines and worn bearing surfaces. Any component found to be worn or damaged must be renewed, repairs are impracticable.

26   Engine/gearbox reassembly: general

1   Before reassembly of the engine/gearbox unit is commenced, the various component parts should be cleaned thoroughly and placed on a sheet of clean paper, close to the working area.

2   Ensure all traces of old gaskets have been removed and the mating
surfaces are clean and undamaged. Great care should be taken when
removing old gasket compound not to damage the mating surface.
Most gasket compounds can be softened using a suitable solvent such
as methylated spirits, acetone or cellulose thinner. The type of solvent
required will depend on the type of compound used. Gasket compound
of the non-hardening type can be removed using a soft brass-wire brush
of the type used for cleaning suede shoes. Do not resort to scraping with
a sharp instrument unless necessary.
3   Gather together all the necessary tools and have available an oil can
filled with clean engine oil. Make sure that all new gaskets and oil seals
are to hand, also all replacement parts required. As a general rule each
moving engine component should be lubricated thoroughly as it is fitted
into position.
4   Make sure that the reassembly area is clean and that there is
adequate working space. Refer to the torque and clearance settings
whenever they are given. Many of the smaller bolts are easily sheared if
overtightened. Always use the correct size screwdriver bit for the
cross-head screws and never an ordinary screwdriver or punch. If the
existing screws show evidence of maltreatment in the past, it is advisable to renew them as a complete set.

27.3 Take input shaft measurement from external faces of 1st and 2nd
gears

27   Reassembling the engine/gearbox unit: rebuilding the gearbox shafts

1   The input and output shaft assemblies should be stored and rebuilt
separately to avoid the accidental interchange of components. Prepare
a clean work surface and have ready a can of fresh gearbox oil so that
each component can be lubricated as it is refitted. Note that the
manufacturer strongly recommends that all circlips be renewed once
disturbed. Take care not to expand the circlips larger than is necessary
to slide them over the shaft. Refer to the accompanying line drawing
and photographs when assembling the various components.
2   Take the bare input shaft with its integral 1st gear, hold the shaft by
its threaded right-hand end and slide the 5th gear over the shaft end so
that its dogs point away from the 1st gear. Fit a retaining circlip to the
groove on the shaft splines to secure the gear. Slide the combined 3rd
and 4th gear onto the shaft with the dogs of the larger 4th gear facing
towards the 5th gear.
3   Thoroughly degrease the shaft shank as far as the start of the
splines and also the 2nd gear boss. Apply Suzuki thread lock 1303B or a
similar locking compound such as Loctite bearing fit to the internal
surface of the pinion and push it as far as possible over the left-hand end
of the shaft and position this against the shaft end and down the shaft
with its plain face outwards. A means must now be found of pressing
the gear onto the shaft to a precise position. While this can be achieved
using a hammer and a long tubular drift to tap the pinion carefully into
place, a better method is as follows. A vice with a jaw opening wide
enough to accept the length of the shaft must be found. Place a soft
alloy or wooden cover over one of the vice jaws and place the threaded
right-hand end of the shaft against this cover. Slip a socket or tubular
drift of suitable length and diameter over the left-hand end of the shaft
and position this against the other vice jaw. With the socket or tubular
drift bearing only on the surface of the 2nd gear, tighten the vice until the
outer faces of the 1st and 2nd gear pinions are precisely 84.5 – 84.6 mm
(3.329 – 3.333 in) apart. This measurement must be checked frequently
using a micrometer or vernier gauge.
4   When the 2nd gear pinion has been pressed home to its correct
fitted position, check that all bearing surfaces have been lubricated and
put the complete input shaft assembly to one side.
5   Take the bare output shaft. Holding its left-hand end slide the 2nd
gear onto it with its flat face butting up against the shaft shoulder and
secure it with a circlip. Fit the sliding dog, larger face first, down the shaft
followed by a circlip. A splined thrust washer is fitted against the circlip
and on top of this, the 3rd gear with its cutouts facing towards the
sliding dog. Fit the 4th gear with its flat face towards the 3rd gear,
followed by a splined thrust washer and circlip. The 5th gear with its
integral selector fork groove, should then be fitted with its selector fork
groove towards the 4th gear; secure with a circlip. Finally fit the 1st gear
on the shaft with its flat face outwards, followed by the plain thrust
washer.

28   Reassembling the engine/gearbox unit: preparing the crank-cases

1   At this stage the crankcase castings should be clean and dry with
any damage, such as worn threads, repaired. If any bearings are to be
refitted, the crankcase casting must be heated first as described in
Section 15.
2   Place the heated casting on a wooden surface, fully supported
around the bearing housing. Position the bearing on the casting, ensuring that it is absolutely square to its housing, then tap it fully into place
using a hammer and a tubular drift such as a socket spanner which bears
only on the bearing outer race. Be careful to ensure that the bearing is
kept absolutely square to its housing at all times.
3   Oil seals are fitted into a cold casing in a similar manner. Apply a thin
smear of grease to the seal circumference to aid the task, then tap the
seal into its housing using a hammer and a tubular drift which bears only
on the hard outer edge of the seal, thus avoiding any risk of the seal
being distorted. Tap each seal into place until its flat outer surface is just
flush with the surrounding crankcase. Oil seals are fitted with the
spring-loaded lip towards the liquid (or gas) being retained, ie with the
manufacturer's marks or numbers facing outwards. Where double-
lipped seals are employed, eg the right-hand main bearing, use the
marks or numbers to position each seal correctly if no notes were made
on removal.
4   Note that the two crankcase main bearing oil seals must have a
thread-locking compound applied to their outer edges before insertion
otherwise they will tend to rotate out of position during use.
5   Where retaining plates are employed to secure bearings or oil seals,
thoroughly degrease the threads of the mounting screws, apply a few
drops of thread-locking compound to them, and tighten them securely.
Note that the input shaft right-hand bearing retainer screws are of two
different head designs. Ensure that the countersunk screw is fitted at
the rear, and the pan-head screw is fitted at the front.
6   When all bearings and oil seals have been fitted and secured, lightly
lubricate the bearings with clean engine oil and apply a thin smear of
grease to the sealing lips of each seal.
7   Support the appropriate crankcase half on two wooden blocks
placed on the work surface; there must be sufficient clearance to permit
the crankshaft and gearbox components to be fitted. Remember that
these are to be fitted into the crankcase left- hand half.

29   Reassembling the engine/gearbox unit: refitting the crankshaft and gearbox components

1   Lightly oil the surface of the crankshaft left-hand end and carefully
ease the crankshaft into the left-hand main bearing, taking care not to

27.5a Fit the 2nd gear to the output shaft and secure with circlip

27.5b Slide the sliding dog over the shaft ...

27.5c ... and secure with the circlip

27.5d Fit a splined thrust washer against the circlip

27.5e Fit the 3rd gear as shown ...

27.5f ... followed by the 4th gear ...

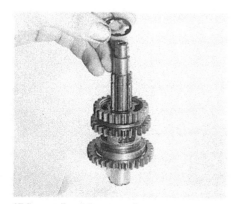

27.5g ... splined thrust washer ...

27.5h ... and circlip

27.5i Fit the 5th gear as shown ...

27.5j ... and secure with the circlip

27.5k Fit the 1st gear ...

27.5l ... and the thrust washer

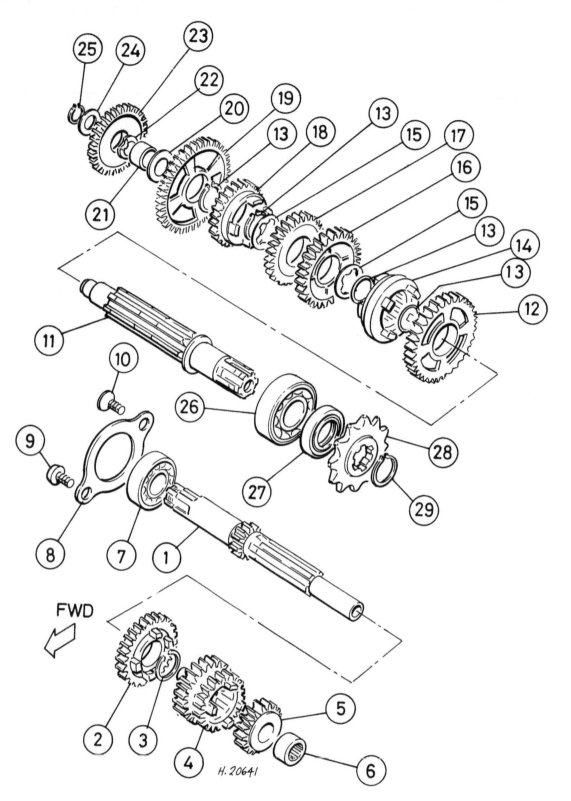

Fig. 1.9 Gearbox shafts

1	Input shaft	9	Pan-head screw	16	Output shaft 3rd gear	23	Kickstart idler pinion
2	Input shaft 5th gear	10	Countersunk-head screw	17	Output shaft 4th gear	24	Washer
3	Circlip	11	Output shaft	18	Output shaft 5th gear	25	Circlip
4	Input shaft 4th/3rd gear	12	Output shaft 2nd gear	19	Output shaft 1st gear	26	Output shaft left-hand bearing
5	Input shaft 2nd gear	13	Circlip – 4 off	20	Thrust washer	27	Oil seal
6	Input shaft left-hand bearing	14	Sliding dog	21	Output shaft right-hand bearing	28	Gearbox sprocket
7	Input shaft right-hand bearing	15	Splined thrust washer – 2 off	22	Wave washer	29	Circlip
8	Bearing retainer plate						

28.4 Apply thread-locking compound to seal's outer edges on fitting

28.5 Refit bearing retaining plates – note different screw types on input shaft bearing retainer

29.1 Insert crankshaft into its bearing in left-hand crankcase

damage the lips of the oil seal; ensure that it is kept vertical and enters the bearing squarely. Refit the primary drive gear nut to the crankshaft, threading the nut down until it is flush with the crankshaft end. Using a soft-faced mallet, tap lightly but firmly on the crankshaft end to drive the crankshaft into place.

2 Excess force will not be required to complete this task; a few gentle

taps should suffice. If it is found that excessive force is needed, it is recommended that the two components be taken to an authorized Suzuki dealer for the work to be undertaken using Suzuki's own service tools.

3 With the crankshaft fitted remove the primary drive pinion nut and check that the crankshaft is free to rotate.

4 Insert the output shaft into its bearing in the casing, taking care that the splines on its end do not damage the oil seal as they pass through; tape can be placed around the splines to prevent oil seal damage. Lift off the thrust washer and 1st gear pinion temporarily to allow the input shaft to be fitted. Fit the input shaft into its bearing and carefully mesh the gears with those of the output shaft. It should be possible to fit the gear clusters by hand alone, but a few light taps on each shaft right-hand end with a soft-faced mallet may be necessary to drive the shafts fully into place. Check that both shafts are free to rotate and refit the 1st gear pinion and thrust washer to the output shaft.

5 Insert the plastic oil pump drive pinion, lubricating its bearing surface with oil.

6 Refit the selector forks to their respective grooves in the gearbox shafts using the notes made on dismantling as a reference to their exact location. If no notes were made, refer to the accompanying line drawings and photographs as a guide to their correct positions. Place the stopper arm in position in the left- hand crankcase, but do not fit its return spring at this stage.

7 Check that the camplate and locating pin are correctly aligned and seated on the selector drum left-hand end and insert the drum into the crankcase, noting that the two small lobes on the camplate must engage the roller of the stopper arm. This will place the selector drum in the neutral position.

8 Slide the selector fork locating pins into their respective tracks in the selector drum and refit both selector fork shafts with the longest shaft fitted to the output shaft forks. Make sure that these shafts are seated

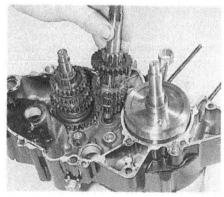

29.4a Insert output shaft and remove its 1st gear while input shaft is fitted

29.4b Ensure shaft gears mesh correctly and refit output shaft 1st gear and thrust washer

29.5 Refit the oil pump drive pinion

29.6a Insert the selector forks in their correct shaft grooves

29.6b Place the stopper arm in position in the casing

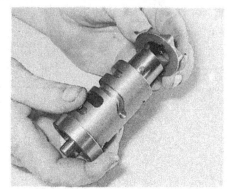

29.7a Fit the camplate over the selector drum end so that it aligns with the pin

29.7b Insert drum into casing so that stopper arm roller aligns with neutral position on camplate

29.8a Position each fork pin in its respective track in the drum ...

29.8b ... and slide the fork shafts into position

correctly into their housings in the crankcase left-hand half and that the shaft retaining the input shaft forks also passes through the stopper arm.

9   Using a pair of long nosed pliers, carefully refit the return spring to the stopper arm and the spring post set in the crankcase.

10   Finally, check that all gearbox components are in place and check that the gear shafts rotate easily in their bearings without any sign of binding. Lubricate all bearing surfaces thoroughly and rotate the selec-

tor drum to ensure that all gears can be selected easily, then return the drum to the neutral position.

30   Reassembling the engine/gearbox unit: joining the crankcase halves

1   Thoroughly degrease the crankcase mating surfaces of both halves and apply a thin film of jointing compound evenly to these surfaces. If the two dowel pins were previously removed these should now be pressed firmly into their locations in the left-hand casing. Ensure that the connecting rod is aligned with the crankcase mouth. Make a final check that all components are lubricated. Note that a smear of grease on the crankshaft right-hand end will greatly assist reassembly.

2   Lower the right-hand crankcase half into position, using firm hand pressure only to push it home. It may be necessary to give a few gentle taps with a soft-faced mallet to drive the casing fully into place. Do not use excessive force, instead check that all shafts and dowels are correctly fitted and accurately aligned, and that the crankcase halves are exactly square to each other. If necessary, pull away the right-hand crankcase half to rectify the problem before starting again.

3   When the two halves have joined correctly and without strain, invert the complete assembly taking care to keep the crankcases together and refit the crankcase retaining screws using the cardboard template to position each correctly. Working in a diagonal sequence from the centre outwards, progressively tighten the screws until all are securely and evenly fastened.

4   Wipe away any excess compound from around the joint area, then check the free running and operation of the crankshaft and gearbox components. If a particular shaft is stiff to rotate, a smart tap on each end using a soft-faced mallet will centralise the shaft in its bearing. If this does not work, or if any other problem is encountered, the crankcases must be separated again to find and rectify the fault.

29.9 Reconnect the stopper arm return spring

30.2a Lower the right-hand crankcase half into position ...

30.2b ... ensuring that the dowel pins are correctly aligned

30.3 Secure the crankcase screws, taking care to position any cable or wiring ties correctly

31 Reassembling the engine/gearbox unit: refitting the gearchange shaft and selector ratchet

1 The selector ratchet must be handled carefully or the pawls will fly out under spring tension. If the components have been separated, refit the springs, rollers and pawls into the ratchet making sure that the largest shoulder on each pawl faces outwards (photo).
2 Squeeze the pawls together against spring tension and fit the assembly into the selector drum end, so that the centre tooth of the ratchet points towards the gearchange shaft bore in the casing (photo).
3 Refit the pawl guide and ratchet locating plates to the crankcase and apply a thread-locking compound to their threads; tighten the screws securely.
4 Check that the spacer and return spring are correctly fitted to the head of the gearchange shaft and push the shaft through the crank-cases, whilst taking care not to damage the oil seal lips as the shaft splines pass through. Note that it is a good idea to wrap a layer of tape around the shaft splines to prevent damage to the oil seal during installation.
5 Fit the gearchange shaft return spring over the spring post in the crankcase so that the post is positioned between the two spring ends. Mesh the gearchange shaft teeth with those of the ratchet so that the ratchet's centre tooth engages between the two centre teeth of the gearchange shaft.
6 Temporarily refit the gearchange lever and check that all gears can be selected with relative ease whilst turning the exposed part of the

output shaft by hand. Check also that the return spring is functioning correctly and brings the gear lever back to its original position after each gear selection is made.

32 Reassembling the engine/gearbox unit: refitting the idler gear and kickstart assembly

1 Fit the wave washer followed by the idler gear pinion, flat face outwards, onto the exposed output shaft right-hand end, then fit the flat washer and secure the assembly with a circlip.
2 Oil the bearing surface of the kickstart shaft and push the shaft into its locating hole set in the crankcase, ensuring that the ratchet stop on the shaft butts against the lug on the crankcase.
3 Ensure that the spring clip is correctly fitted in the groove on the kickstart drive gear pinion. Slide the pinion over the kickstart shaft end and push it into place whilst rotating the clip so that it seats correctly between the two ribs cast in the crankcase. Fit the spring support washer over the shaft end. Slide the kickstart return spring over the shaft and insert its inner end into the hole provided in the kickstart shaft. Slide the plastic spring guide over the shaft so that its slot engages the spring inner end.
4 Rotate the spring clockwise until it can be hooked over the spring post set in the crankcase.

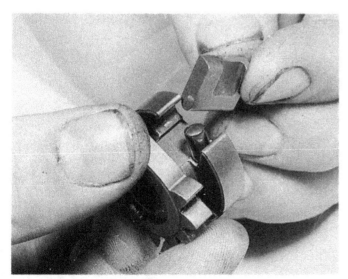

31.1 Assemble the springs, plungers and pawls in the ratchet, noting that the pawls' largest shoulders must face outwards

31.2 While holding the pawls in position insert the ratchet into the selector drum

31.3 Refit the pawl guide and ratchet locating plates, applying a locking compound to their threads

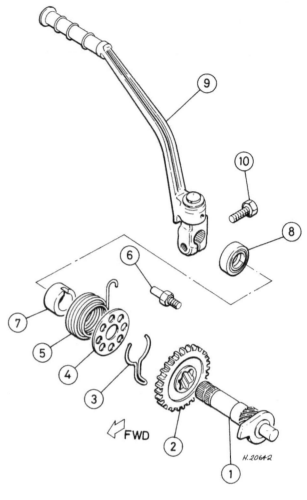

Fig. 1.10 Kickstart mechanism

1	Kickstart shaft	6	Spring post
2	Drive gear	7	Spring guide
3	Spring clip	8	Oil seal
4	Spring support washer	9	Kickstart lever
5	Return spring	10	Pinch bolt

31.5 Position the return spring ends each side of the gearchange lever post

32.1a Fit the wave washer and kickstart idler gear to the output shaft end ...

32.1b ... fit the washer and secure with the circlip

32.2 Kickstart shaft stop must butt against crankcase lug (arrowed)

32.3a Slide kickstart pinion, complete with spring clip, over shaft. Engage clip end in casing

32.3b Install the spring support washer ...

32.3c ... the return spring and its guide

32.4 Hook the return spring end over the post

bush and the outer drum. Note that it may be necessary to rotate the shafts slightly to allow the teeth of the outer drum to mesh with those of the primary drive pinion and kickstart idler gear. Fit the second thrust washer and the clutch centre.

5   Employ the same method of preventing the clutch centre from turning as used in the dismantling procedure. Use a new tab washer, then fit and tighten the nut to the correct torque setting. Flatten part of the tab washer against one of the flats of the nut.

6   Lubricate the long pushrod and insert it into the input shaft, followed by the short headed pushrod and thrust bearing; ensure that all are well lubricated.

7   The three friction and two plain metal plates can now be fitted. If new plates are being fitted, they must first be coated with a light film of transmission oil. Start with a friction plate followed by a plain plate, building up the plates alternately.

8   Fit the pressure plate to the clutch assembly and insert the springs, washers and bolts. Tighten the bolts evenly and in a diagonal sequence to the specified torque setting.

9   If dismantled, the release mechanism should now be reassembled, greasing the operating worm lightly, and refitted to the crankcase left-hand cover.

33 Reassembling the engine/gearbox unit: refitting the primary drive pinion and clutch

1   Fit the spacer over the right-hand crankshaft end with its stepped shoulder facing outwards and insert the Woodruff key into the keyway.

2   Slide the primary drive pinion onto the shaft with its stepped shoulder facing outwards and align its slot with the Woodruff key. The lock washer is of the Belville type and must be fitted with its convex surface facing outwards.

3   Lock the crankshaft by the same method used for its removal, fit the nut and tighten to the specified torque setting.

4   Place the thrust washer over the input shaft end, followed by the

34 Reassembling the engine/gearbox unit: refitting the right- hand crankcase cover

1   Check that all components of the clutch, primary drive, kickstart and gear selector mechanism are correctly fitted. If either the tachometer drive or driven gears have been removed they must be refitted as described in Chapter 4, Section 17. Refit the crankcase cover locating dowels and fit a new gasket to the mating surface of the crankcase, sticking it in place with a smear of grease if necessary.

2   Ensure that the teeth of the tachometer drive pinion mesh correctly with the teeth of the primary drive pinion. This can be assisted by slowly rotating the crankshaft as the crankcase cover is being fitted. Take care not to damage the oil seal lips as the kickstart shaft splines pass through the cover.

33.1 Slide the spacer over the crankshaft right-hand end and insert the Woodruff key

33.2 Fit the primary drive pinion as shown and fit the lock washer

33.3 Install the retaining nut and lock the engine while the nut is tightened

33.4a Insert the thrust washer and bush over the input shaft end ...

33.4b ... followed by the outer drum and second thrust washer

33.5a Install the clutch centre, tab washer and nut ...

33.5b ... then tighten the nut while holding the centre as shown

33.5c Flatten a section of the tab washer against the nut flats

33.6 Insert the long and short pushrods and thrust bearing

33.7a Fit the friction ...

33.7b ... and plain plates alternately

Install the pressure plate, springs, washers and bolts

3   Insert the retaining screws into their correct positions. Tighten them evenly, working from the centre outwards and in a diagonal sequence to avoid distorting the casting.

35   Reassembling the engine/gearbox unit: refitting the neutral indicator switch, gearbox sprocket and oil pump

1   Insert the spring followed by the contact pin into the hole in the left-hand end of the selector drum. Renew the O-ring and refit the neutral switch cover, securing it with the two screws.
2   Use a new paper gasket to seal the joint between the oil pump and crankcase mating surfaces. Ensure that the tang on the back of the pump engages correctly with the slot in the oil pump drive gear when it is fitted. Refit the spring washers, and tighten the two pump retaining screws securely.
3   Slide the gearbox sprocket onto the output shaft and secure it with the circlip. Note that if there is any doubt about the condition of the circlip it must be renewed.

36   Reassembling the engine/gearbox unit: refitting the flywheel generator

1   Refit the stator plate to its original position by aligning the marks made prior to dismantling (see Section 7) and securing it to the left-hand crankcase with the three screws.
2   If no marks were made on dismantling, the initial setting can be achieved by aligning the right-hand stator plate mark with the crankcase index mark (see photo 7.5). The timing must then be checked with a stroboscope as described in Chapter 3 on completion of the rebuild.
3   Refit the grommet and secure the generator lead by its clamp to the crankcase ensuring that it is correctly routed well away from any moving parts. Reconnect the neutral switch lead.
4   Degrease the rotor and crankshaft mating surfaces and remove any metal particles or swarf from the rotor magnets. Insert the Woodruff key into the crankshaft and push the rotor over it, aligning its cutout with the key. Lock the crankshaft using one of the methods described in Section 7 and tighten the nut to the specified torque setting, having first applied a thread-locking compound to its threads.

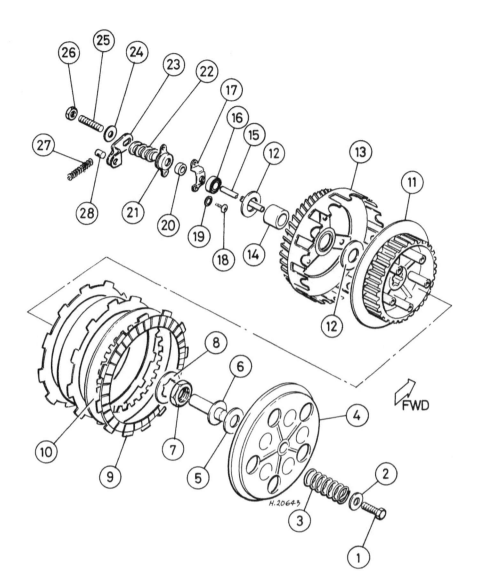

Fig. 1.11 Clutch

1   Bolt – 5 off
2   Washer – 5 off
3   Spring – 5 off
4   Pressure plate
5   Thrust bearing
6   Headed pushrod
7   Nut
8   Tab washer
9   Friction plate – 3 off
10 Plain plate – 2 off
11 Clutch centre
12 Thrust washer – 2 off
13 Outer drum
14 Bush
15 Long pushrod
16 Oil seal
17 Retaining plate
18 Screw – 2 off
19 Washer – 2 off
20 Dust seal
21 Release body
22 Operating worm
23 Operating lever
24 Washer
25 Adjusting screw
26 Locknut
27 Spring
28 Cable trunnion

34.1a Refit the dowel pins in the crankcase surface ...

34.1b ... and place a new gasket over them before refitting the cover

35.1 Use a new O-ring at the neutral switch
joint

35.2 Use a new gasket at the oil pump mating
surface

35.3 Slide gearbox sprocket onto output shaft
and secure with circlip

36.1 Refit the stator plate, aligning the marks
made on dismantling

36.3 Reconnect the neutral switch lead

36.4a Insert Woodruff key in crankshaft end ...

36.4b ... and refit rotor. Use holding tool as shown while nut is
tightened

---

37 Reassembling the engine/gearbox unit: refitting the piston, cylinder barrel and head

---

1   Position the engine/gearbox unit so that it is upright on the work
surface. Rotate the crankshaft to raise the connecting rod to its highest
point and thoroughly lubricate the big-end bearing with clean engine oil.
Wipe clean the crankcase to cylinder barrel mating surface before
easing the new base gasket into position over the barrel retaining studs.
Take care to fit the base gasket the correct way over the studs; do not

risk damaging the gasket by attempting to force it into position. This
point applies equally to the cylinder head gasket. Pack the crankcase
mouth with clean rag in order to prevent any component parts from
falling into the crankcase during the following fitting procedures. Refit
the two cylinder barrel locating dowels into their locations in the
crankcase surface.
2   Lubricate the small-end eye of the connecting rod and the small-end
bearing itself with clean engine oil before pushing the bearing into
position. Place the piston over the connecting rod so that the arrow cast
in the piston crown faces forward and slide the gudgeon pin into
position. The pin should be a light sliding fit but if it proves to be tight,
warm the piston in hot water to expand the metal around the gudgeon
pin bosses. Use new circlips to retain the gudgeon pin, and double check
to ensure that each is correctly located in the piston boss groove. If a
circlip works loose, it will cause serious engine damage. The circlips
should be fitted so that the gap between the circlip ends is well away
from the cutout to the side of the gudgeon pin hole. Finally, check that
the piston rings have not been disturbed from the positions quoted in
Section 22 of this Chapter.
3   Prior to refitting the barrel check the reed valve assembly to ensure
that its mountings are secure and that it is clean. Lubricate both the
cylinder bore and piston rings with clean engine oil. Position two blocks
of wood across the crankcase mouth, one each side of the connecting
rod, and carefully lower the piston onto the blocks. This will provide
positive support to the piston whilst easing the rings into the bore.
4   Place the cylinder barrel in position over its retaining studs and
proceed to lower it carefully down over the piston. Guide the piston
crown into the bore and push in on each side of the piston rings so that
they slide into the bore. There is a generous lead in the base of the bore
which will aid this operation. Take care that the ring ends stay each side
of the ring pegs; if the rings ride up over the pegs breakage is certain.
With the rings safely inserted into the cylinder bore, remove the blocks
from underneath the piston and the rag from the crankcase mouth. Push
the cylinder barrel firmly down onto the crankcase and tap it lightly
around its upper surface with a soft-faced mallet to ensure that it is
properly seated. Replace the two cylinder barrel retaining nuts on their

37.1 Refit dowel pins and place a new gasket over holding studs

37.2a Lubricate and fit small-end bearing

37.2b Install piston and insert gudgeon pin. Note arrow marking on piston crown must face forwards

37.2c Secure gudgeon pin with new circlips. Note rag packing crankcase mouth

37.4 Lower barrel carefully over piston

37.5a Fit a new cylinder head gasket over the studs ...

studs and tighten them by hand only at this stage. Unplug the oil feed pipe and push it back over the brass union. Slide the clamp down the length of the feed pipe until it is firmly seated over the union.

5   Clean the mating surfaces of both the cylinder barrel and the cylinder head. Carefully fit the head gasket, noting that there is again only one way it will fit correctly. Lower the cylinder head into position and fit, then tighten, the four nuts. The nuts should be tightened in several stages in a diagonal sequence to the specified torque setting. Tighten the two cylinder barrel nuts to the specified torque setting.

37.5b ... install the head and tighten the nuts in a diagonal sequence

38   Refitting the engine/gearbox unit into the frame

1   Fitting the engine/gearbox unit into the frame is a straightforward reversal of the removal procedure. Take a few minutes to remove any corrosion from the engine mounting bolts and clean up their threads. In the case of the rear mounting, which is formed by the swinging arm pivot bolt, grease the bolt shank. Note that the nuts used to fasten the engine mounting bolts are of the nylon insert self-locking type and should be renewed whenever they are disturbed.

2   Ease the engine/gearbox unit into its correct location in the frame taking care not to damage the paintwork or any of the alloy engine casings. Align the engine with its mountings and push the two mounting bolts and the swinging arm pivot bolt through from left to right. Fit the washer to the swinging arm pivot bolt and fit a new locknut to all three bolts, tightening them to the specified torque setting.

3   Slide the kickstart lever onto its splined shaft aligning the marks made on dismantling to return it to its original position; refit and tighten the pinch bolt.

4   Refit the tachometer cable into its housing in the right- hand engine casing. It may be necessary to rotate the crankshaft by gently pressing down on the kickstart in order to align the tangs on the cable with the slots in the cable drive housing. Tighten the knurled ring to secure the cable.

5   Refit the carburettor to the intake stub, using new gaskets on each side of the insulating spacer. Refit the two washers and nuts, tightening them securely. Insert the throttle slide into the carburettor body ensuring that the slot in the slide aligns correctly with the projection cast in the carburettor body and that the needle enters its jet easily. Tighten the carburettor top fully.

6   Route the oil pump cable through the adjuster set in the crankcase top surface and connect it to the pump operating lever, using the nylon holder in the lever end. Working as described in Routine maintenance, check the throttle cable/oil pump cable adjustment, and make any adjustment necessary. Using a spout type oil can filled with the recommended engine oil prime the oil feed pipe (pump to cylinder barrel) and reconnect the free end to the pump union, securing the pipe with the spring clip. Taking care to avoid as little oil loss as possible, reconnect

the oil feed pipe to the oil tank. Replenish the oil tank if necessary and then bleed the oil feed side of the pump as described in Chapter 2, Section 14. Do not refit the pump cover at this stage; the oil injection system must be bled of air with the engine running on completion of rebuild.

7   Use a smear of grease to stick a new gasket to the exhaust port. Refit the exhaust pipe and silencer tightening their mountings by hand. Tighten the exhaust port fasteners first then the remaining mountings to the specified torque settings.

8   Route the main generator lead from the crankcase up to its main loom connector on the frame, remembering to secure it to the frame with the tie provided. Refit the spark plug, having cleaned and reset its gap, and connect the spark plug cap.

9   If the battery was removed it can now be refitted. Remake the terminal connections, remembering to connect the negative (–) terminal last. Take care to route the battery vent pipe correctly so that it is not trapped and its end is placed well clear of the frame tubes. Refit the battery retaining band.

10   Refit the drive chain (closed end of the spring link facing the normal direction of chain travel) and adjust if necessary the chain tension, rear brake adjustment and stop lamp switch height referring to Routine maintenance for information.

11   Note that prior to refitting the crankcase left-hand cover it will be necessary to bleed the oil pump and check the ignition timing. Since both operations require the engine to be running these are discussed in the following section.

12   Refit the fuel tank, checking there is no metal-to-metal contact which will split the tank. Reconnect the fuel feed pipe, turn on the tap and check for leaks, which must be cured immediately. Refit the frame front covers.

13   Refit the seat and sidepanels, check that the transmission oil drain plug is refitted and tightened to the specified torque setting, then refill the gearbox with oil to the correct level as described in Routine maintenance. Note, however, that slightly more oil may be required to reach the correct level after a full engine rebuild; refer to the Specifications Section of Chapter 2.

14   Make a final check that all components have been refitted and that all are securely fastened and correctly adjusted.

39   Starting and running the rebuilt engine

1   Start the engine using the procedure for a cold engine. A certain amount of perseverance may prove necessary to coax the engine into activity even if new parts have not been fitted. Should the engine persist in not starting, check that the spark plug has not become fouled by the oil used during reassembly. Failing this, go through the fault diagnosis section and work out what the problem is methodically.

2   When the engine does start, keep it running as slowly as possible to allow the oil to circulate. Open the choke as soon as the engine will run without it. During the initial running, a certain amount of smoke may be in evidence due to the oil used in the reassembly sequence being burnt away. The resulting smoke should gradually subside. As soon as the engine will run smoothly, carry out the procedure described in Chapter 2 to bleed air from the oil injection system. On completion of the bleeding operation refit the pump cover. Referring to Chapter 3, Section 7, check the ignition timing and make any adjustment necessary. When both operations are complete, refit the crankcase left-hand cover, securing it with the four screws. If the clutch cable was previously disconnected, this must be passed through the adjuster set in the cover and connected to the clutch operating arm trunnion. Adjust the clutch cable as described in Routine maintenance and refit the inspection cover. Refit the gear lever to its shaft, noting that the alignment mark on the shaft end should correspond with the clamp opening.

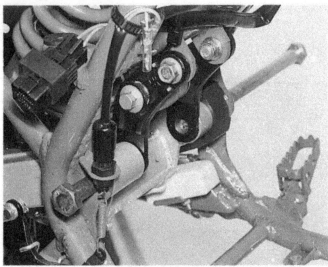

38.1 Prior to refitting the engine support the rear suspension with two bolts as shown

3   Check the engine for blowing gaskets and oil leaks. Before using the machine on the road, check that all the gears select properly, and that the controls function correctly.

40   Taking the rebuilt machine on the road

1   Any rebuilt machine will need time to settle down, even if parts have been replaced in their original order. For this reason it is highly advisable to treat the machine gently for the first few miles to ensure oil has circulated throughout the lubrication system and that any new parts fitted have begun to bed down.

2   Even greater care is necessary if the engine has been rebored or if a new crankshaft has been fitted. In the case of a rebore, the engine will have to be run in again, as if the machine were new. This means greater use of the gearbox and a restraining hand on the throttle until at least 500 miles have been covered. There is no point in keeping to any set speed limit; the main requirement is to keep a light loading on the engine and to gradually work up performance until the 500 mile mark is reached. These recommendations can be lessened to an extent when only a new crankshaft is fitted. Experience is the best guide since it is easy to tell when an engine is running freely.

3   Remember that a good seal between the piston and the cylinder barrel is essential for the correct functioning of the engine. A rebored two-stroke engine requires careful running-in. There is a far greater risk of engine seizure during the first hundred miles if the engine is permitted to work hard.

4   If at any time a lubrication failure is suspected, stop the engine immediately and investigate the cause. If an engine is run without oil, even for a short period, irreparable engine damage is inevitable.

5   Do not on any account add oil to the petrol under the mistaken belief that a little extra oil will improve the engine lubrication. Apart from creating excess smoke, the addition of oil will make the mixture much weaker, with the consequent risk of overheating and engine seizure. The oil pump alone should provide full engine lubrication.

6   When the initial run has been completed allow the engine unit to cool and then check all the fittings and fasteners for security. Re-adjust any controls which may have settled down during initial use and check the transmission oil level, topping up if necessary.

Chapter 2 Fuel system and lubrication

Contents

Specifications

Fuel tank
Total capacity .... 7.0 lit (1.54 gal)
Reserve... 1.3 lit (2.29 Imp pint)

Fuel grade .... Unleaded or low-lead, minimum 85 octane (RON/RM)

Carburettor
Make .... Mikuni
Type VM15SH
Bore.. 15 mm
ID number .... 13620 (XKE to XKM), 13690 (XKR)
Main jet 67.5
Air jet .... 1.0 mm
Jet needle.. 3DJ01
Clip position – grooves from top................................... 4th
Needle jet E-7
Throttle valve cutaway 2.0
Pilot jet... 17.5
Pilot outlet.. 0.7 mm
Valve seat... 1.2 mm
Starter jet 35
Pilot screw – turns out 1½
Float height .... 22.4 ± 1.0 mm (0.88 ± 0.04 in)
Idle speed .... 1400 ± 100 rpm

Air filter
Type .... Oiled polyurethane foam

Engine lubrication system
Type .... Suzuki CCI pump feed total-loss system
Tank capacity.. 1.2 litres (2.1 Imp pint)
Oil grade .... Suzuki CCI, CCI Super or a good quality two-stroke oil suitable for oil injection systems

Gearbox lubrication system
Type .... Oil bath
Capacity:
    At oil change... 850 cc (1.5 Imp pint)
    At engine rebuild .... 950 cc (1.67 Imp pint)
    Oil grade .... SAE 20W/40 motor oil

Torque wrench settings

Component	kgf m	lbf ft
Reed valve petal screws	0.7 – 0.9	5 – 7
Exhaust pipe to barrel bolts	0.4 – 0.7	2.9 – 5.0
Exhaust pipe to frame bolt	1.0 – 1.6	7 – 12

1 General description

Fuel from the frame-mounted tank is gravity fed to the float chamber of the Mikuni carburettor via a three position fuel tap. Air is drawn into the carburettor through an air cleaner casing, housing an oil impregnated polyurethane foam filter. The proportions of air and atomized fuel are regulated by the carburettor, which provides the correct mixture at all throttle positions. Cold starting is assisted by a manual choke, located on the carburettor right-hand side.

Engine lubrication is by a pump-fed total-loss system known as Suzuki CCI. Two-stroke oil is gravity fed from a frame-mounted tank to the oil pump. A cable connecting the oil pump and throttle twistgrip ensures delivery of the correct quantity of oil as the engine speed varies. Gearbox and primary drive components are lubricated by splash from a supply of oil in the reservoir formed by the crankcase castings.

The exhaust system is a two-part assembly comprising the exhaust pipe and expansion chamber with a separate silencer fitted to the end.

2 Fuel tank: removal, examination and refitting

Note: *Petrol is extremely flammable, especially in the form of vapour. Take all precautions to prevent the risk of fire and read the Safety first! section of this manual before starting work.*

1   If it is necessary to remove the fuel tank for repairs the following points should be noted. Fuel tank repair, whether necessitated by accident damage or by fuel leaks, is a task for the professional. Welding or brazing is not recommended unless the tank is purged of all fuel vapour; which is a difficult condition to achieve. Resin-based tank sealing compounds are a much more satisfactory method of curing leaks, and are now available through suppliers who advertise regularly in the motorcycle press. Accident damage repairs will inevitably involve re-painting the tank; matching of modern paint finishes, especially metallic ones, is a very difficult task not to be lightly undertaken by the average owner. It is therefore recommended that the tank be removed by the owner, and then taken to a motorcycle dealer or similar expert for

professional attention.

2   Carefully detach the sidepanels, frame front covers and remove the dualseat, then turn the petrol tap to the Off position. Use a suitable pair of pliers to release the wire petrol pipe retaining clip to allow the pipe to be pulled off the stub at the rear of the tap. Careful use of a small screwdriver may be necessary to help ease the pipe off the stub. Once the pipe is detached, allow any fuel in the pipe to drain into a small clean container. The tank may now be detached from the frame by removing the retaining bolt at the rear of the tank and pulling it up and rearwards off its front mounting rubbers. Place the tank mounting components in a safe place ready for refitting. Inspect the mounting rubbers for signs of damage or deterioration and if necessary renew them before refitting of the tank is due to take place.

3   To refit the tank, reverse the removal procedure. Move it from side to side before it is fully home, so that the rubber buffers engage with the guide channels correctly. If difficulty is encountered in engaging the front of the tank with the rubber buffers, apply a small amount of lubricant to the buffers to ease location. Secure the tank with the retaining bolt whilst ensuring that the mounting components are correctly located and that there is no metal to metal contact between the tank and frame.

4   Finally, always carry out a leak check on the fuel pipe connections after fitting the tank and turning the tap lever to the On position. Any leaks found must be cured; as well as wasting fuel, any petrol dropping onto the hot engine castings presents a real fire risk.

3 Fuel tap: removal, examination and refitting

Note: *Petrol is extremely flammable, especially in the form of vapour. Take all precautions to prevent the risk of fire and read the Safety first! section of this manual before starting work.*

1   The fuel tap assembly is secured to the underside of the tank by two bolts, the joint being sealed by an O-ring.

2   If work is being carried out to cure fuel leaks, first ensure that all retaining fasteners are tight but are not overtightened. The tap components are delicate and easily distorted; overtightening will merely

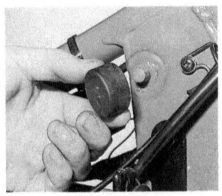
2.2a Tank locates in mounting rubbers at front ...

2.2b ... and is retained by single bolt at rear

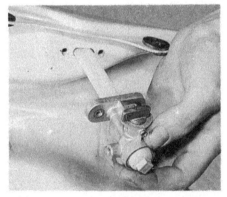
3.1 Fuel tap is retained to tank by two bolts

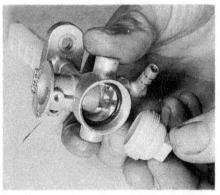
3.4a Unscrew the sediment bowl and pick out the O-ring

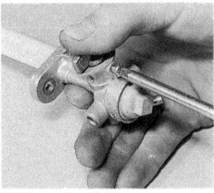
3.4b Remove the grub screw ...

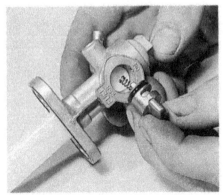
3.4c ... to permit removal of the tap lever

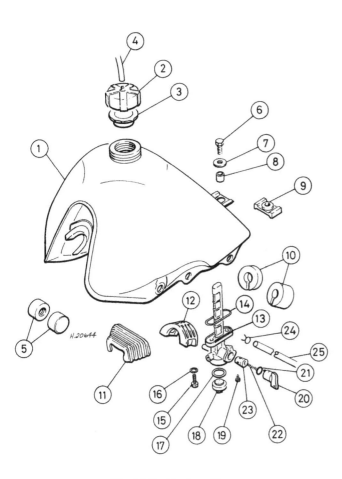

Fig. 2.1 Fuel tank and tap

1	Fuel tank	14	O-ring
2	Filler cap	15	Bolt – 2 off
3	Gasket	16	Washer – 2 off
4	Vent pipe	17	O-ring
5	Front mounting rubber – 2 off	18	Sediment bowl
6	Bolt	19	Grub screw
7	Washer	20	Tap lever
8	Spacer	21	O-ring
9	Damping rubber	22	Spring
10	Damping rubber – 2 off	23	Valve
11	Damping rubber	24	Clip
12	Damping rubber	25	Fuel pipe
13	Fuel tap		

exacerbate any leaks. If the components are correctly fastened and the leaks persist, the tap must be stripped and the relevant seal renewed.

3   Petrol tap removal must be preceded by draining the petrol remaining in the tank. Remove the petrol feed pipe from the carburettor, switch the tap to the Res position and allow all the fuel to drain into a clean, dry, container of suitable size. Note that this container must be suitable for holding petrol and clearly marked if the petrol is to be stored for any length of time. Alternatively, the tank can be removed as described in the preceding Section, and placed on one side so that the petrol level is below that of the tap. Take care not to damage the tank paintwork if this method is employed, and ensure that petrol does not leak from the filler cap breather pipe.

4   Unscrew the sediment bowl by using a close-fitting ring spanner on the moulded square end of the bowl. Pick out the sealing O-ring. Remove the two bolts which retain the tap to the underside of the tank and carefully withdraw the tap assembly. The filter gauze can then be cleaned in new petrol, using a soft-bristled brush to remove any deposits. The tap lever is retained by a grub screw threaded into the underside of the tap body. Remove the screw, then withdraw the tap

lever. It may be necessary to pad the jaws of a suitable pair of pliers with cloth or tape and to pull the lever out as it is sealed by a tight-fitting O-ring. Pick out the small spring and the shaped plug behind it.

5   Thoroughly clean and inspect all the tap components, renewing any parts that are worn or damaged. Again, pay particular attention to the fibre sealing washers beneath the heads of the two tap mounting bolts and to the large O-ring set in the tap mating flange surface. It is recommended that these components are renewed whenever the tap is disturbed as their condition is critical to the prevention of persistent petrol leaks.

4   Petrol feed pipe: examination

Note: *Petrol is extremely flammable, especially in the form of vapour. Take all precautions to prevent the risk of fire and read the Safety first! section of this manual before starting work.*

1   The petrol pipe is made from thin-walled synthetic rubber and is of the push-on type, secured by a retaining clip at both ends. The manufacturer recommends that the fuel feed pipe be renewed every four years as a matter of course, although if cracking or splitting is noticed the pipe should be renewed immediately.

2   If the pipe has been replaced with a transparent plastic type for any reason, look for signs of yellowing which indicate that the pipe is becoming brittle due to the plasticiser being leached out by the petrol; if apparent the pipe should be renewed with the correct synthetic rubber type. On no account should natural rubber tubing be used to carry petrol, even as a temporary measure. The petrol will dissolve the inner wall, causing blockages in the carburettor jets which will prove very difficult to move.

5   Carburettor: removal and refitting

Note: *Petrol is extremely flammable, especially in the form of vapour. Take all precautions to prevent the risk of fire and read the Safety first! section of this manual before starting work.*

1   Removal of the carburettor is a straightforward task requiring no preliminary dismantling.

2   Turn the fuel tap to the Off position and disconnect the fuel pipe at the carburettor union, catching any surplus petrol in a clean container for subsequent re-use. Slacken and remove the float chamber drain plug. If the presence of dirt or water is suspected in the carburettor, catch the petrol which drains out in a clean container. This will reveal traces of whatever foreign matter is causing the problem. Unscrew the carburettor top and withdraw the throttle valve assembly from the carburettor,

5.2 Carburettor front and rear mountings (arrowed)

tape the throttle cable to the frame to keep the delicate throttle valve and needle assembly out of harm's way. Slacken the air filter hose clamp, withdraw the hose, and then slacken and remove the two carburettor mounting nuts and their respective washers. Manoeuvre the carburettor backwards off its studs and clear of the machine. If the insulating spacer is disturbed, it should be removed, cleaned and refitted with new gaskets to prevent induction leaks.

3   Fitting the carburettor is a straightforward reversal of the removal procedure, whilst noting the following points. When refitting the throttle valve assembly into the carburettor body, take great care to ensure that the jet needle enters the needle jet smoothly and correctly and that the slot cut in the side of the throttle valve is in correct alignment with the projection cast in the carburettor body.

4   Do not overtighten the two carburettor retaining nuts or the carburettor body may be distorted, resulting in a loss of performance due to induction leaks. Adjust the throttle cable at the adjuster on the carburettor top to give 0.5 – 1.0 mm (0.02 – 0.04 in) of free play, measured in terms of twistgrip rotation. Note that throttle cable adjustment should only be carried out after the carburettor has been correctly set and that the oil pump cable adjustment must be checked whenever the throttle cable setting is altered. Refer to Routine maintenance for details of oil pump cable adjustment.

6   Carburettor: overhaul

Note: *Petrol is extremely flammable, especially in the form of vapour. Take all precautions to prevent the risk of fire and read the Safety first! section of this manual before starting work.*

1   Before dismantling the carburettor, cover an area of the work surface with clean paper or rag. This will not only prevent any components that are placed upon it from becoming contaminated with dirt, moisture or grit but, by making them more visible, will also prevent the many small components removed from the carburettor body from becoming lost.

2   Proceed to dismantle the carburettor by removing the four screws and spring washers that retain the float chamber to the main body of the carburettor. Note that it may be necessary to tap around the float chamber joint with a soft-faced mallet to break its seal.

3   With the float chamber removed and placed to one side, pull the pivot pin from the twin float assembly and lift the floats from position. The float needle can now be displaced from its seating and should be put aside in a safe place for examination at a later stage.

4   Unscrew and remove the single crosshead screw which serves to retain the float needle seat retaining plate in position between the two float pivot pin columns. Withdraw the plate and pull the needle seat out of its location in the carburettor body.

5   Unscrew and remove the main jet. Note that, when unscrewing any jet from the carburettor, a close fitting screwdriver of the correct type must be used to prevent damage occurring to the soft material from which the jet is constructed. With the main jet removed, the needle jet may be lifted out of the carburettor body. Note the gasket fitted to its end. Unscrew and remove the pilot jet.

6   Note the setting of the throttle stop screw by counting the number of turns required to screw it fully in. Remove the throttle stop screw, taking care to retain its spring. Note the setting of and remove the pilot screw with its spring. Failure to note the settings of the aforementioned screws will make it less easy to 'retune' the carburettor after it has been reassembled and refitted to the machine.

7   The choke plunger valve and spring can be withdrawn after the retaining nut has been unscrewed and the operating lever disconnected.

8   The only removable component fitted to the float chamber of the carburettor is the drain plug which takes the form of a single slotted screw with a sealing washer located beneath its head. It is not necessary to remove this plug except for renewal of the sealing washer or renewal of the screw itself.

9   Prior to examination of the carburettor component parts, clean each part thoroughly in clean petrol before placing it on a piece of clean rag or paper. Use a soft nylon-bristled brush to remove any stubborn contamination on the castings and blow dry each part with a jet of compressed air. Avoid using a piece of rag for cleaning since there is always risk of particles of lint obstructing the airways or jet orifices. Never use a piece

6.2 Remove four retaining screws to release float chamber

6.3a Pull out float pivot pin and lift off float assembly

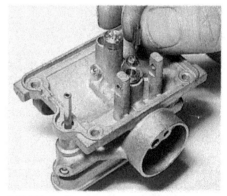
6.3b Lift the float needle out of its seating

6.4 Remove the single screw and retaining plate to release float needle seat

6.5a Unscrew and remove the main jet ...

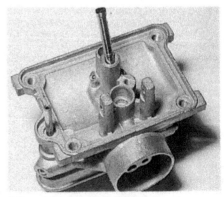
6.5b ... then lift out the needle jet

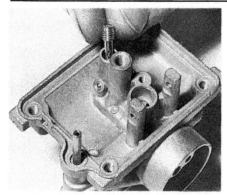

6.5c Pilot jet is located next to main jet column

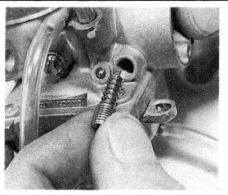

6.6 Record their exact positions before removing pilot or throttle stop screws

6.7 Inspect the choke plunger valve for wear

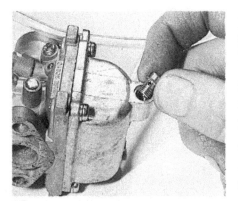

6.8 Drain plug is located in base of float chamber

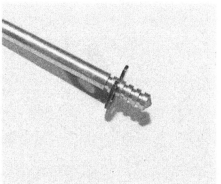

6.14a Check that jet needle is straight and clip is in correct groove

6.14b When fitting jet needle to throttle valve position retaining clip as shown

of wire or any pointed metal object to clear a blocked jet, it is only too easy to enlarge a jet under these circumstances and increase the rate of petrol consumption. If an air line is not available, a blast of air from a tyre pump will usually suffice. If all else fails to clear a blocked jet, remove a bristle from the soft-bristled brush and carefully pass it through the jet to clear the blockage.

10   Check each casting for cracks or damage and check that each mating surface is flat by laying a straightedge along its length. Any distorted casting must be renewed.

11   Remove all O-rings and sealing gaskets from the component parts and replace them with new items. Ensure that, where applicable, they are correctly seated in their retaining grooves. The springs on the throttle stop and pilot screws should now be carefully inspected for signs of corrosion and fatigue and renewed if necessary.

12   The seating area of the float needle will wear after lengthy service and should be closely examined with a magnifying glass. Wear usually takes the form of a ridge or groove, which will cause the float needle to seat imperfectly. Any wear of this nature will necessitate renewal of the needle and seat.

13   Closely examine the twin float assembly for signs of damage, especially around the soldered joints. Shake the float to establish if a leak is present. Although it is theoretically possible to repair a brass float by soldering, any attempt to do so is likely to cause a small but dramatic explosion, having a detrimental effect on both the float and the operator. It is far preferable to renew the float, although a safer temporary repair may be made with Araldite or Petseal.

14   Inspect the throttle valve for wear. This wear will be denoted by polished areas on the external diameter. Excessive wear will allow air leaks, weakening the mixture, and producing erratic slow running. Many mysterious carburation maladies may be attributed to this defect, the only cure being to renew the valve, and if worn badly in corresponding areas, the carburettor body. If removal of the valve is necessary, grasp the valve firmly in one hand whilst compressing the return spring against the carburettor top with the other. Disengage the throttle cable from its retaining slot in the valve and withdraw the retaining spring clip,

followed by the jet needle and its clip, from the valve. Before renewing the valve, take care to inspect each individual component part, commencing with the needle.

15   Examine the needle carefully for scratches or wear along its length. Ensure that the needle is not bent by rolling it on a flat surface, such as a sheet of plate glass. If in doubt as to the condition of the needle, return it to an authorized Suzuki dealer who will be able to give further advice and, if necessary, provide a new component.

16   Inspect the return spring for signs of fatigue, failure or severe corrosion and renew it if found necessary. The procedure adopted for reassembly of the throttle valve component parts should be a direct reversal of that used for dismantling.

17   Inspect the surface of the choke valve for signs of scratches or excessive wear and the return spring for signs of fatigue, failure or corrosion. Note that the choke components are not available separately, and if renewal is required the complete carburettor body must be renewed. However, it may be found that the same part is fitted to another model in the range and is available separately. Consult your local Suzuki dealer for his help, should this problem arise.

18   Prior to reassembly of the carburettor, check that all the component parts, both new and old, are clean and laid out on a piece of clean rag or paper in a logical order. On no account use excessive force when reassembling the carburettor because it is easy to shear a jet or some of the smaller screws. Furthermore, the carburettor is cast in a zinc based alloy which itself does not have a high tensile strength. If any of the castings are damaged during reassembly, they will almost certainly have to be renewed.

19   Reassembly is basically a reversal of the dismantling procedure, whilst noting the following points. If in doubt as to the correct fitted position of a component part, refer either to the figure accompanying this text or to the appropriate photograph. When fitting the throttle stop and pilot screws, ensure that each screw is first screwed fully in, until it seats lightly and then unscrewed to its previously noted position. Alternatively, set the pilot screw to the position given in the Specifications Section of this Chapter, as a basis for correct adjustment.

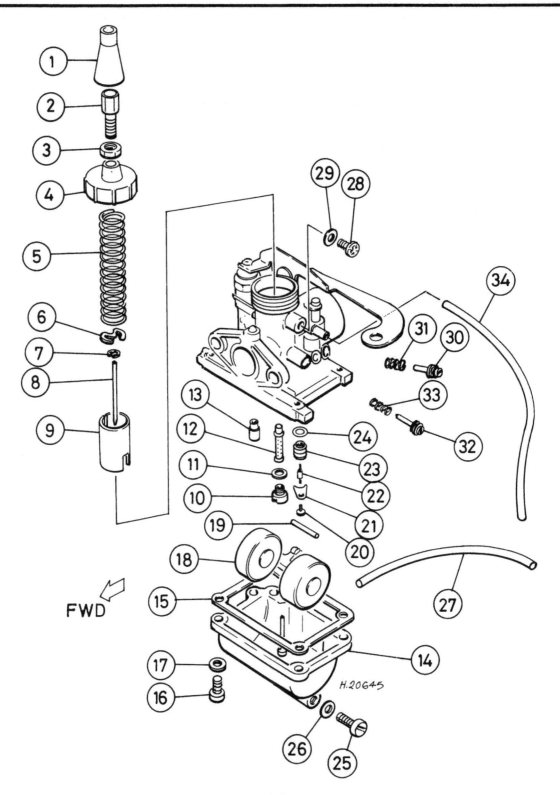

Fig. 2.2 Carburettor

1	Rubber cover	10	Main jet
2	Cable adjuster	11	Washer
3	Locknut	12	Needle jet
4	Carburettor top	13	Pilot jet
5	Return spring	14	Float chamber
6	Needle retainer	15	Gasket
7	Needle clip	16	Screw – 4 off
8	Jet needle	17	Spring washer – 4 off
9	Throttle valve	18	Float

19	Pivot pin
20	Screw
21	Retaining plate
22	Float needle valve
23	Valve seat
24	O-ring
25	Drain screw
26	Sealing washer
27	Overflow pipe

28	Inspection plug
29	Sealing washer
30	Pilot screw
31	Spring
32	Throttle stop screw
33	Spring
34	Breather pipe

7.2 Float height measurement

7.8a Adjust engine idle speed using throttle stop screw

7.8b Pilot air screw adjustment

7 Carburettor: adjustments and settings

Note: *Petrol is extremely flammable, especially in the form of vapour. Take all precautions to prevent the risk of fire and read the Safety first! section of this manual before starting work.*

1   If flooding of the carburettor or excessive mixture weakness have been experienced, it is wise to start operations by checking the float height, which will involve removing the carburettor from the machine and removing the float chamber and gasket.

2   To measure the float level, stand the inverted carburettor on a flat and level work surface and measure the distance between the gasket surface of the carburettor body and the furthest point of the float as shown in the accompanying photograph. The float level is correct when the measurement obtained is that given in the Specifications Section of this Chapter, and is altered by bending very gently and carefully the brass tang which is situated between the two floats and against which the float needle actually bears.

3   The various jet sizes, throttle valve cutaway and needle position are predetermined by the manufacturer and should not require modification. Check with the Specifications list at the beginning of this Chapter if there is any doubt about the types fitted.

4   Before any attempt at adjustment is made, it is important to understand which parts of the instrument control which part of its operating range. A carburettor must be capable of delivering the correct fuel/air ratio for any given engine speed and load. To this end, the throttle valve, or slide as it is often known, controls the volume of air passing through the choke or bore of the instrument. The fuel, on the other hand is regulated by the pilot and main jets, by the jet needle, and to some extent, by the amount of cutaway on the throttle valve.

5   As a rough guide, up to ⅛ throttle is controlled by the pilot jet, ⅛ to ¼ by the throttle valve cutaway, ¼ to ¾ throttle by the needle position and from ¾ to full throttle by the size of the main jet. These are only approximate divisions, which are by no means clear cut. There is a certain amount of overlap between the various stages.

6   If any particular carburation fault has been noted, it is a good idea to try to establish the most likely cause before dismantling or adjusting takes place. If, for example, the engine runs normally at road speeds, but refuses to tick over evenly, the fault probably lies with the pilot system, and will most likely prove to be an obstructed jet. Whatever the problem may appear to be, it is worth checking that the jets are clear and that all the components are of the correct type. Having checked these points, refit the carburettor and check the settings as follows.

7   Refer to Routine maintenance and adjust the throttle cable free play, oil pump cable adjustment and engine idle speed.

8   Set the pilot screw to the position given in the Specifications Section. Start the engine, and allow it to reach its normal working temperature. This is best done by riding the machine for a few miles. Set the throttle stop screw to the specified idling speed. Try turning the pilot screw inwards by about ¼ turn at a time, noting its effect on the idling speed, then repeat the process, this time turning the screw outwards. The pilot screw should be set in the position which gives the fastest consistent tickover. If desired, the tickover speed may be reduced further by lowering the throttle stop screw, but care should be taken that this does not cause the engine to falter and stop after the throttle twistgrip has been opened and closed a few times.

9   Always guard against the possibility of incorrect carburettor adjust-

ment which will result in a weak mixture. Two-stroke engines are very susceptible to this type of fault, causing rapid overheating and often subsequent engine seizure. Changes in carburation leading to a weak mixture will occur if the air cleaner is removed or disconnected, or if the silencer is tampered with in any way. Above all, do not add oil to the petrol, in the mistaken belief that it will aid lubrication. Adequate lubrication is provided by the throttle controlled oil pump.

8 Air filter: general

1   The element must be kept clean and renewed if damaged if good performance is to be maintained. Apart from the obvious problem of the increased wear caused by a damaged element, a clogged or broken filter will upset the mixture setting, allowing it to become too rich or too weak.

2   Unless the air filter is damaged and obviously in need of renewal, it should require no attention other than the regular servicing described in the Routine maintenance section.

3   Never run the engine without the air filter element fitted. The carburettor is set up with the air filter fitted, and if removed, the resulting weak mixture will lead to subsequent engine damage.

9 Reed valve assembly: examination and renovation

1   The reed valve assembly is mounted on the underside of the cylinder barrel, and it follows that access can only be gained by removing the cylinder barrel as described in Chapter 1 Section 6. The reed valve operation is automatic and requires no maintenance, although it can be checked as described below if failure is suspected.

9.2 Reed valve block is secured to barrel by two screws

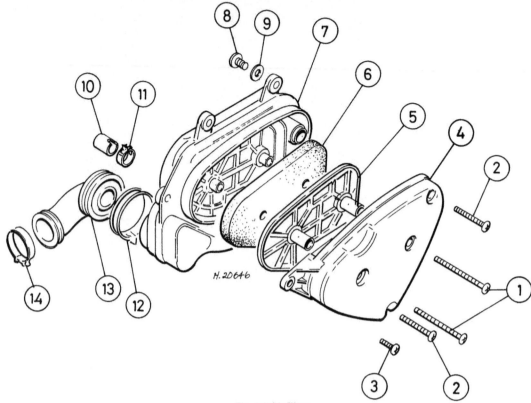

Fig. 2.3 Air filter

1 Screw – 2 off	5 Support frame	9 Washer – 2 off	13 Carburettor hose
2 Screw – 2 off	6 Element	10 Drain pipe	14 Hose clamp
3 Screw	7 Casing	11 Clip	
4 Cover	8 Screw – 2 off	12 Hose clamp	

2   Once the cylinder barrel has been removed, remove the two cross-head screws on the valve case and the reed valve assembly can be lifted away. Note that it is a precision component and can easily be rendered inoperative if dropped or allowed to become dirty.

3   It is not necessary to dismantle the unit to check wear or distortion in the reeds. The distance between the open end of the reed petals and the edge of the valve case port should be greater than 1 mm (0.04 in), see accompanying illustration. If this distance is less, the reed petals can be considered worn and must be replaced. Distortion in the reeds can be measured by the gap between the reed petal and the valve case surface. This must not exceed 0.2 mm (0.008 in) anywhere along the length of the reed petals. If it does, the reed petals must be replaced. Do not attempt to bend the reed petals straight in case fatigue failure is induced. This can cause fragments of broken reed petal to be drawn into

the engine possibly causing severe damage or even complete engine seizure.

4   Examine the condition of the various components for signs of fatigue or cracking and renew if necessary. Should the reed stopper plate and petal retaining screws be removed, a thread-locking compound such as Loctite should be applied to their threads and the screws tightened to the recommended torque setting.

5   Check that there are no traces of dust or grit on the component especially in between the reed petals and stopper plate as this dirt will hinder the opening of the reeds, resulting in poor engine performance.

6   Secure the reed valve assembly to the underside of the barrel, and refit the barrel as described in Chapter 1, Section 37.

10   Exhaust system: removal and refitting

1   The exhaust system is a two-piece arrangement which has a total of four mounting points. Two mountings are obscured by the fuel tank and right-hand sidepanel and it is necessary to remove the seat, sidepanels, frame front covers and fuel tank if removal of the complete system is required. Remove the bolt securing the silencer and carefully pull the silencer backwards clear of the rubber connecting sleeve and main exhaust section. Remove the two front bolts mounting the exhaust to the engine and the bolt and nut securing the exhaust to the frame. Carefully manoeuvre the exhaust system forwards, clear of the engine and withdraw it from the machine taking great care not to damage the paintwork of either the exhaust or the machine. Remove and discard the exhaust port gasket.

2   Inspect all the rubber mounting components and the rubber connecting sleeve and renew any part which is worn or damaged. Apart from the front mounting at the engine, the other three mounting points all have damping rubbers fitted.

3   Refitting is a straightforward reversal of the removal procedure. Always use a new gasket to prevent exhaust leaks, sticking it to the

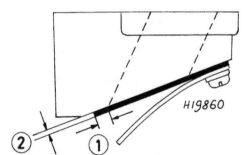

Fig. 2.4 Reed valve measurement points – see text

1   Valve petal edge to casing port measurement
2   Valve petal to casing surface measurement

10.1a Remove silencer mounting bolt and pull silencer free of main exhaust

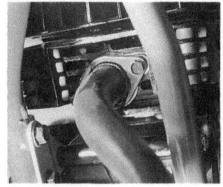

10.1b Slacken and remove front mounting bolts ...

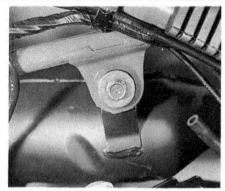

10.1c ... bolt ...

10.1d ... and nut, then withdraw exhaust main section

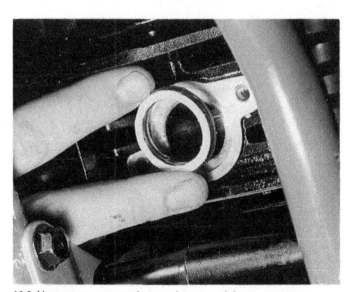

10.3 Always use a new gasket at exhaust port joint

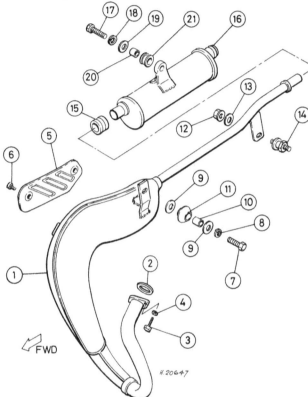

Fig. 2.5 Exhaust system

1 Exhaust pipe	12 Nut
2 Gasket	13 Washer
3 Bolt – 2 off	14 Mounting rubber/stud
4 Spring washer – 2 off	15 Connecting sleeve
5 Heat shield	16 Silencer
6 Screw – 2 off	17 Bolt
7 Bolt	18 Spring washer
8 Spring washer	19 Washer
9 Washer – 2 off	20 Spacer
10 Spacer	21 Mounting rubber
11 Mounting rubber	

---

11 Exhaust system: decarbonisation and renovation

1   Due to the fact that the exhaust system has no removable baffles, exhaust cleaning will be rather more difficult than in most machines of this type.

2   The exhaust gases of any two-stroke are particularly oily in nature and will produce a marked build-up of carbon deposits in the front length of the exhaust pipe and of an oily sludge in the rear part of the system. If not removed at regular intervals, these deposits will cause an undesirable increase in back pressure in the system, restricting the engine's ability to breathe and producing a marked loss in power output.

exhaust port with a thin smear of grease. The main section of the assembly must be offered up first and secured in position by refitting loosely all of the mounting bolts. Fit the silencer, ensuring the rubber sleeve is correctly aligned and then tighten all of the mounting bolts securely, to the specified torque setting (where given) working down the length of the system, starting at the front mounting.

3   Three possible methods of cleaning the system exist, the first, which will only be really effective if the deposits are very oily, consists of flushing the system with a petrol/paraffin mixture. It will be evident that great care must be taken to prevent the risk of fire when using this method, and that the system must be hung overnight in such a way that all the mixture will drain away. It will be necessary to use a suitable scraping tool to remove any hardened deposits from the front length of the exhaust pipe and from the exhaust port of the cylinder barrel.

4   The second method will involve the use of a welding torch or powerful blow lamp to burn off the deposits. This usually results in the production of a great deal of smoke and fumes and so must be carried out in a well-ventilated area. It also requires some considerable skill in the use of welding equipment if it is to be fully effective, and if personal injury or damage to the exhaust system is to be avoided. Another drawback is that the excessive heat created will destroy the painted finish of the system, and that repainting will be necessary. With severely blocked systems, it may even be necessary to cut the exhaust open so that the flame can be applied to the blocked section, and the exhaust welded up afterwards. In short this method, while quite effective, should only be employed by a person with welding equipment and the necessary degree of skill in its use. The best solution would be for the owner to remove the exhaust from the machine and to take it to a local dealer or similar expert for the work to be carried out.

5   The third possible method of exhaust cleaning is to use a solution such as caustic soda to dissolve the deposits. While this is a lengthy and time-consuming operation, it is the simplest and most effective method that can be used by the average owner, and is therefore described in detail in the following paragraphs. Bear in mind that it is very important to take great care when using caustic soda as it is a very dangerous chemical. Always wear protective clothing, this must include proper eye protection. If the solution does come into contact with the eyes or skin it must be washed clear immediately with clean, fresh, running water. In the case of an eye becoming contaminated, seek expert medical advice immediately. Also, the solution must not be allowed to come into contact with aluminium alloy – especially at the above recommended strength – caustic soda reacts violently with aluminium and will cause severe damage to the component. To clean the silencer casing and exhaust pipe assembly, remove the system from the machine and suspend it from its silencer end.

6   Block up the end of the exhaust pipe with a cork or wooden bung. If wood is used, allow an outside projection of three or four inches with which to grasp the bung for removal. The mixture used is a ratio of 3 lbs caustic soda to a gallon of fresh water. This is the strongest solution ever likely to be required. Obviously the weaker the mixture the longer the time required for the carbon to be dissolved. Note, whilst mixing the solution, that the caustic soda should be added to the water gradually, whilst stirring. **Never** pour water into a container of caustic soda powder or crystals; this will cause a violent reaction to take place which will result in great danger to one's person.

7   Commence the cleaning operation by pouring the solution into the system until is quite full. Do not plug the open end of the system. The solution should now be left overnight for its dissolving action to take place. Note that the solution will continue to given off noxious fumes throughout its dissolving process; the system must therefore be placed in a well ventilated area. After the required time has passed, carefully pour out the solution and flush the system through with clean fresh water. The cleaning operation is now complete.

8   When cleaning the exhaust system do not forget to remove any carbon deposits from the exhaust port itself. Use a blunt edged scraping tool of suitable shape and size, and take great care not to damage the piston or to scratch the surface of the port. If the port is severely blocked, the cylinder barrel should be removed for a major decarbonising operation.

9   Do not at any time attempt to modify the exhaust system in any way. The exhaust system is designed to give the maximum power possible consistent with legal requirements and yet to produce the minimum noise level possible. Quite apart from the legal aspects of attempting to modify the exhaust it is very unlikely that an unskilled person could improve the performance of the machine by working on the exhaust. If an aftermarket accessory system is being considered, check very carefully that it will maintain or increase performance when compared with the standard system. Very few 'performance' exhaust systems live up to the claims made by the manufacturers, and even fewer offer any performance increase at all over the standard component.

10   The final point to be borne in mind when considering the exhaust

system is the finish. The matt-black painted finish employed in pursuit of current styling trends is cheaper to renovate but less durable than a conventional chrome-plated system. It is inevitable that the original finish will deteriorate to the point where the system must be removed from the machine and repainted, therefore some thought must be given to the type of paint to be used. As it is such a common problem, especially amongst the owners of modern trail bikes, several alternative finishes are offered. Reference to the advertisements in the national motorcycle press, or to an authorized Suzuki dealer and to the owners of machines with similarly-finished exhausts will help in selecting the most effective finish. The best are those which require the paint to be baked on, although some aerosol sprays are almost as effective. Whichever finish is decided upon, ensure that the surface is properly prepared according to the paint manufacturer's instructions and that the paint itself is correctly applied.

12   Oil tank: removal and refitting

1   Unlock and raise the dualseat. Remove the right-hand sidepanel.
2   Lift out the tool tray and disconnect the oil level sender unit wires.
3   The oil feed pipe should now be disconnected from the base of the tank, accessible from the right-hand side of the machine. Before discon-

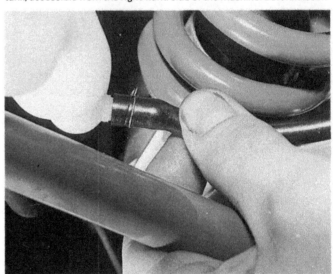

12.3 Have ready a prepared piece of pipe before oil feed pipe is disconnected from tank union

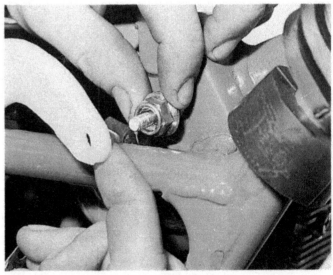

12.4 Suspension unit top bolt doubles as tank front mounting remove first nut and ease tank extension off bolt

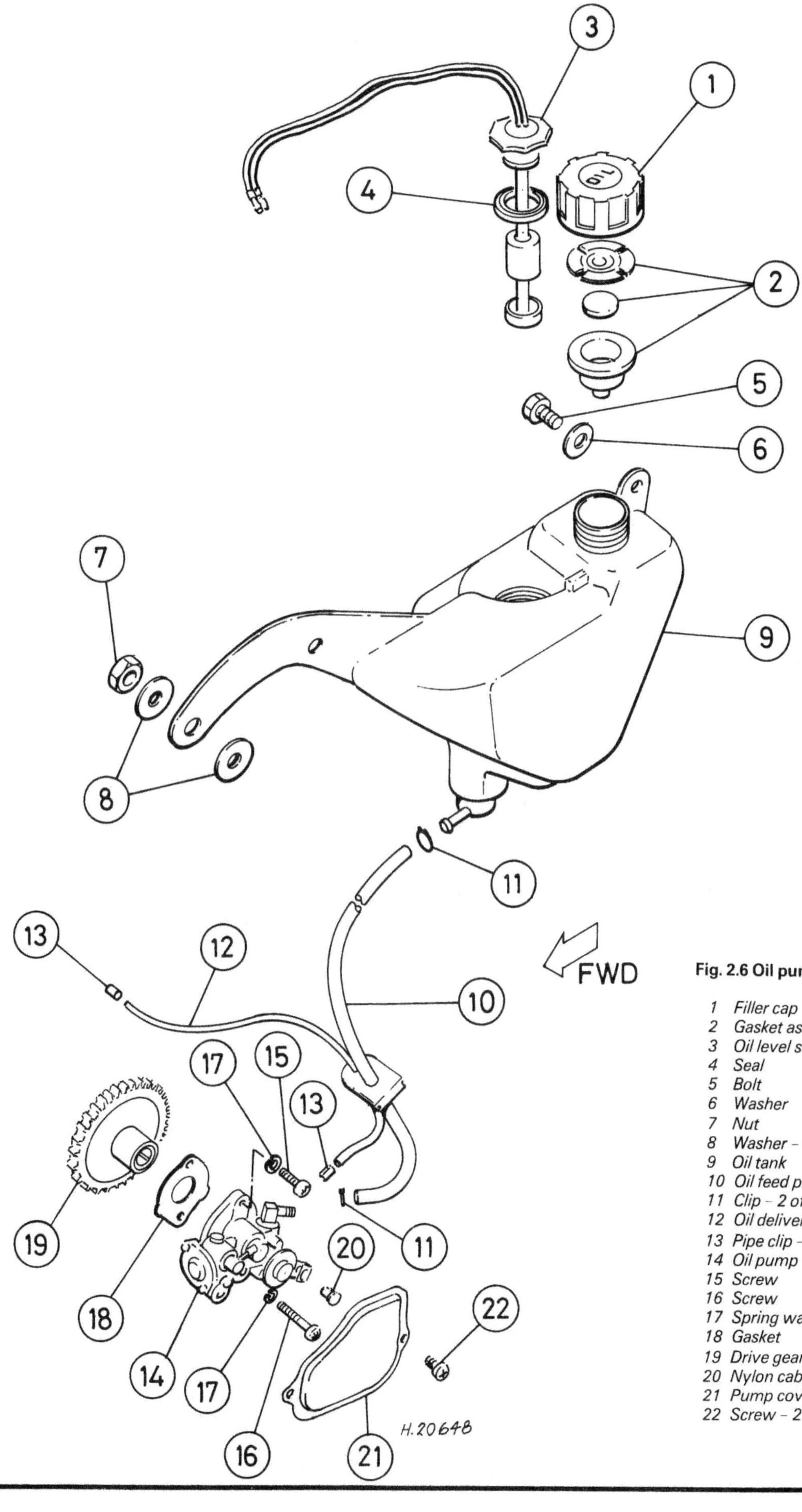

FWD

Fig. 2.6 Oil pump and tank

1. Filler cap
2. Gasket assembly
3. Oil level sender unit
4. Seal
5. Bolt
6. Washer
7. Nut
8. Washer – 2 off
9. Oil tank
10. Oil feed pipe
11. Clip – 2 off
12. Oil delivery pipe
13. Pipe clip – 2 off
14. Oil pump
15. Screw
16. Screw
17. Spring washer – 2 off
18. Gasket
19. Drive gear
20. Nylon cable holder
21. Pump cover
22. Screw – 2 off

H.20648

necting the pipe prepare a short length of pipe, of the same internal diameter, which is plugged at one end. Also have ready a clean screw of suitable size to plug the end of the existing pipe once disconnected. These simple measures will prevent the loss of oil and seal the oil feed pipe while disconnected. Slide the feed pipe clip downwards, off the tank stub. Very swiftly pull the pipe off the tank stub and replace it with the prepared pipe which is closed at one end. Use the screw to plug the oil feed pipe.

4   The tank is secured to the frame by a bolt and nut which once removed will allow the tank to be withdrawn from the top of the machine.

5   Details of the oil level sender unit circuit can be found in Chapter 6, Section 9.

6   Refitting is a reversal of the removal procedure. Refit the tank mounting washers in their original positions and reconnect the oil lever sender unit wires correctly. Ensure that the oil feed pipe is secured to the tank stub by its clip.

7   Bleed the oil feed side of the injection system as described in Section 14.

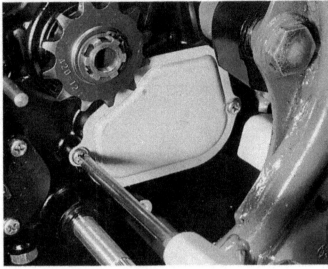

13.3 Remove inspection cover to gain access to oil pump

---

13   Oil pump: removal and refitting

1   The oil pump can be expected to give long service, requiring no maintenance, but in the event of failure it must be renewed. No replacement parts are obtainable and the pump is, therefore effectively a sealed unit. While test procedures are given to measure the pump's delivery rate, these involve the use of test equipment only likely to be available to a Suzuki dealer. It is therefore recommended that the machine be taken to an authorised Suzuki dealer for the pump's output to be checked if doubts arise as to its efficiency. Some idea of the problems involved in checking the pump's output may be gained by the fact that with the pump control lever in the fully open position and with the engine rotating at the recommended test speed of 4000 rpm, the pump will deliver only 0.8 – 1.0 cc of oil over a two minute period.

2   If it is necessary to remove the pump assembly for any reason, slacken and remove the four screws which retain the crankcase left-hand cover, withdraw the cover and manoeuvre it forward to avoid removal of the clutch cable from its mechanism. Disconnect the chain at its connecting link, lift the chain off the teeth of the gearbox sprocket, and hang the chain over the swinging arm pivot.

3   Remove the two screws that serve to retain the oil pump cover plate in position and manoeuvre the cover clear of the machine. Disconnect the pump control cable from the pump lever by pushing the end of the lever up so that tension is taken off the cable inner and then detach the cable nipple from its nylon holder. With the rubber sealing cap detached from the cable adjuster, the cable may now be pulled through the adjuster so that it is clear of the pump.

4   It is now necessary to make provision for catching any oil that will issue from the feed and delivery pipes once they are disconnected from the pump. To prevent complete draining of the oil tank, the feed pipe should be plugged as soon as it is disconnected; a clean screw or bolt of the appropriate thread diameter is ideal for this purpose. Slacken and remove both of the cross-headed retaining screws and lift the pump unit clear of its driveshaft end. Discard the pump base gasket and replace it with a new item.

5   To refit the pump, clean both the pump and crankcase mating surfaces, place the new gasket onto the pump mating surface and align the central driven spigot of the pump with the slot in the end of its drive gear before fitting the pump into its crankcase housing. With the pump correctly seated, fit and tighten to its two retaining screws. Note that each one of these screws must have a serviceable spring washer located beneath its head.

6   Unplug and reconnect both pipes. Ensure that both pipes are a good push fit on their respective stubs and that the large diameter pipe is correctly retained by its spring clip. Check the pipes for any signs of splitting or perishing and ensure that all connections in the lubrication system are free from leaks. Any fault found must be rectified immediately, as leakage will cause a loss of lubrication and possibly complete engine seizure. Route both pipes correctly, taking care to ensure that they are neither twisted nor crimped between any component parts.

7   The oil pump must now be bled of air by following the instructions listed in the following Section of this Chapter.

8   Connect the pump control cable to the nylon holder in the pump

13.6 On refitting, ensure oil feed and delivery pipes are secured by their retaining clips

lever end; check the pump cable adjustment as described in Routine maintenance.

9   Thread the chain around the gearbox sprocket and connect it again. Ensure that the spring clip of the connecting link is correctly fitted, ie with its closed end in the direction of travel of the chain, and that the chain tension is correctly adjusted as described in Routine maintenance.

10   Refit the crankcase right-hand cover, ensuring that the clutch pushrod locates correctly with its release mechanism. Insert the retaining screws and tighten them in a diagonal sequence to avoid damaging the casing. Check the clutch operation and adjust the cable if necessary.

---

14   Oil pump: bleeding of air

1   If the oil pump has been removed, or the oil system drained, then it is most important to bleed any air from the system before the engine is started. Failure to do this will result in the engine seizing, with the resulting expense of a complete engine rebuild and the potential danger to the rider of the machine, should the rear wheel lock unexpectedly

whilst the machine is in motion.

2 Bleeding is effected by removing the cross-headed screw on the front of the pump body and waiting for a steady, air-free, stream of oil to emerge before retightening the screw.

3 If the presence of air is suspected in the delivery side of the pump, ie in the oil pump/cylinder barrel feed pipe, it must be removed by forcing oil through the pipe. This can be done by filling an oil can with the recommended oil, inserting the oil can spout into the pump end of the feed pipe, and by pushing oil through the pipe with the pumping action of the can. Alternatively, the same task can be carried out by running the engine at a steady tickover whilst holding the pump control lever fully open with a finger to increase the delivery rate as much as possible. This latter method has the disadvantage of taking a long time, about 10–15 minutes, and of running the risk of damaging the engine due to over-heating or due to running with an interrupted supply of oil. For this reason, the engine speed must be kept as low as possible and an alternative method of lubrication must be created by substituting the petrol in the tank for a small quantity (1 pint would be ample) of petrol/oil mixture at 25–50:1 ratio, to provide lubrication by the old-fashioned petroil method. The mixture must be drained from the petrol tank after the bleeding operation is completed and the original, clean, petrol replaced in the tank.

4 Fortunately, if sufficient care is taken to plug oil feed pipes as soon as they are disconnected, the above procedure will rarely be necessary.

14.2 Oil pump bleed screw location

15 Gearbox lubrication: general

1 General maintenance for the gearbox lubrication system consists solely of checking the level of oil within the gearbox casing at frequent intervals and changing the oil at the specified interval. Carrying out these two service procedures will preclude any risk of the gearbox components becoming starved of oil or having to run in oil that has deteriorated to the point where it is ineffective in its prime function as a lubricating medium.

2 Full details of checking the oil level and of carrying out an oil change can be found in Routine maintenance.

Chapter 3 Ignition system

Contents

Specifications

Ignition system.. CDI Capacitor Discharge Ignition

Ignition timing.. 18 ± 2° BTDC @ 6000 rpm

Ignition source coil
    Resistance – black/red to black/white .... 120 – 190 ohm

Ignition HT coil/CDI unit
    Primary windings resistance – black/red to coil mounting .... See text
    Secondary windings resistance – plug cap to coil mounting .... 14 – 17 K ohm

Spark plug
    Type .... NGK BP6ES or ND W20EP
    Gap .... 0.6 – 0.8 mm (0.024 – 0.032 in)

1 General description

A simple form of capacitor discharge ignition is employed which requires little or no maintenance. Apart from the ignition switch and related wiring, the system comprises only the combined ignition HT coil/CDI unit, this being supplied with current and triggered at the correct moment by a source coil built into the generator stator.

2 Ignition system: fault diagnosis

1 With the exception of ignition timing adjustment, no provision for adjustment exists, and thus any failure of the system can be traced to the breakdown of a system component or a simple wiring fault. Of the two possibilities, the latter is by far the most likely. In the event of failure, check the system in a logical fashion, as described below.

2 Remove the spark plug, giving it a quick visual check, noting any obvious signs of flooding or oiling. Fit the plug into the plug cap and rest it on the cylinder head so that the metal body of the plug is in good contact with the cylinder head metal. The electrode end of the plug should be positioned so that sparking can be checked as the engine is spun over.

3 **Note:** *The energy levels in electronic ignition systems can be very high. On no account should the ignition be switched on whilst the plug or plug cap are being held. Shocks from the HT circuit can be most unpleasant. Secondly, it is vital that the plug is in position and soundly earthed when the system is checked for sparking. The ignition components can be seriously damaged if the HT circuit becomes isolated.*

4 Having observed the above precautions, turn the ignition switch to 'On' and kick the engine over. If the system is in good condition a regular, fat blue spark should be evident at the plug electrodes. If the spark appears thin or yellowish, or is non-existent, further investigation will be necessary. Before proceeding further, turn the ignition off and remove the key as a safety measure.

5 Ignition faults can be divided into two categories, namely those where the ignition system has failed completely, and those which are due to a partial failure. The likely faults are listed below, starting with the most probable sources of failure. Work through the list systematically, referring to the subsequent sections for full details of the necessary checks and tests.

    (a)   Defective spark plug or spark plug cap
    (b)   Loose, corroded or damaged wiring connections, broken or shorted wiring between any of the ignition system components
    (c)   Faulty ignition or engine kill switch
    (d)   Faulty source coil
    (e)   Faulty ignition HT coil/CDI unit

3 Ignition system: checking the wiring

1 The wiring should be checked visually, noting any signs of corrosion around the various terminals and connectors. If the fault has developed in wet conditions, it follows that water may have entered any of the connectors or switches, causing a short circuit. A temporary cure can be effected by spraying the relevant area with one of the proprietary de-watering aerosols, such as WD40 or similar. A more permanent solution is to dismantle the switch or connector and coat the exposed

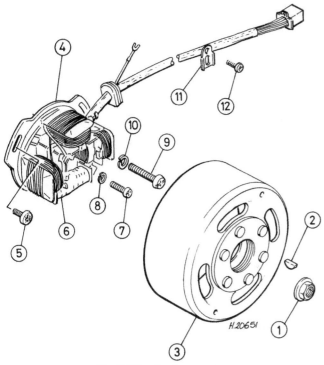

Fig. 3.1 Flywheel generator

1 Nut	7 Screw – 2 off
2 Woodruff key	8 Spring washer – 2 off
3 Rotor	9 Screw – 2 off
4 Stator	10 Spring washer – 2 off
5 Screw – 3 off	11 Cable clamp
6 Ignition source coil	12 Screw

parts with silicone grease to prevent the ingress of water. The exposed backs of connectors can be sealed off using a silicone rubber sealant.

2 Light corrosion can normally be cured by scraping or sanding the affected area, though in serious cases it may prove necessary to renew the switch or connector affected. Check the wiring for chafing or breakage, particularly where it passes close to part of the frame or its fittings. As a temporary measure, damaged insulation can be repaired with PVC tape, but the wire concerned should be renewed at the earliest opportunity.

3 Using the wiring diagram at the end of the manual, check each wire for breakage or short circuits using a multimeter set on the resistance

scale or a dry battery and bulb wired as shown in Fig. 6.1. In each case, there should be continuity between the ends of each wire.

4 Ignition and engine kill switches: testing

1 Using the test equipment described in the previous Section and the wiring diagram at the back of this Manual, check whether the switch connections are being made and broken as indicated in the diagram. Do not forget to check the engine kill switch when checking for faults in the ignition system.

2 Note that it will be necessary to remove the headlamp cowling to gain access to the switch connectors. If the switch is found to be defective, usually renewal will be required, although nothing can be lost by attempting a repair.

3 The ignition switch is secured via a bracket to the steering top yoke and is fastened by the handlebar retaining bolts. The kill switch is retained to the right-hand handlebar end by two screws.

5 Ignition source coil: location and testing

1 The ignition source coil is mounted on the generator stator plate. It can be tested in place, after removing the seat and separating the generator lead from its multi-pin connection, but if removal is required the left-hand crankcase cover and flywheel generator must be withdrawn as described in Section 7 of Chapter 1. Refitting is described in Section 36 of the same Chapter.

2 To test the coil, use a good quality multimeter set to the ohms scale; disconnect the generator lead at its snap connector and connect the meter positive (+) probe to the black/red wire terminal. Connect the meter negative (–) probe either to the black/white wire terminal or to a suitable earth point on the crankcase. The reading obtained should be the same as that given in the Specifications at the beginning of this Chapter.

3 If the reading obtained does not correspond with that given, the coil must be considered faulty and renewed, although it is worthwhile taking the machine to an authorized Suzuki dealer for your findings to be confirmed. Check also that the fault is not due to a trapped or broken wire which can be repaired easily.

6 CDI unit and ignition HT coil: location and testing

1 The CDI unit and ignition coil are combined in a single sealed unit mounted on the frame top tube by two screws. The entire ignition system, with the exception of the ignition source coil in the flywheel generator, is contained in the one unit. It follows that in the unlikely

4.3 Ignition switch is secured to top yoke by handlebar clamp bolts

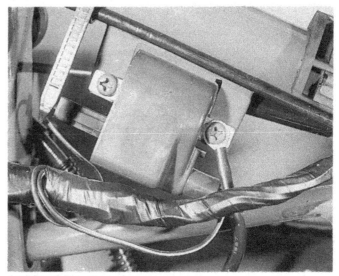

6.1 CDI unit/HT coil location

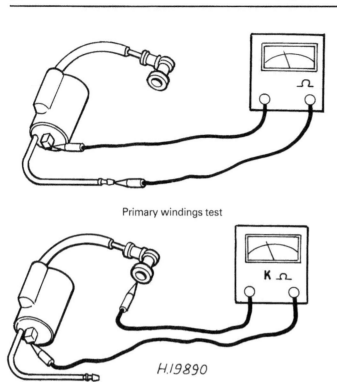

Primary windings test

Secondary windings test

H.19890

Fig. 3.2 Ignition HT coil windings test connections

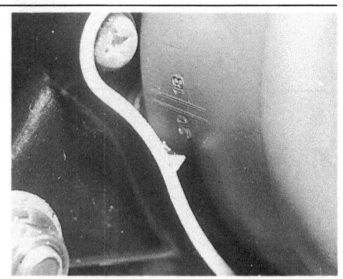

7.2 Upper timing mark on rotor (18) should align with pointer if timing is correct

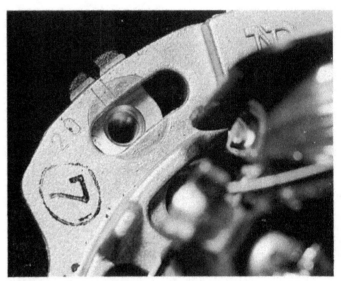

7.4 Timing adjustment is made by slackening stator screws and rotating plate (screw removed for clarity)

event of either component failing, the unit must be renewed complete, repairs are not practicable.

2   In the event of a suspected fault, the primary and secondary coil windings can be tested using a multimeter, although testing on a spark gap tester will give a more accurate indication of the unit's condition. Remove the unit from the machine, noting that this will necessitate removal of the fuel tank and frame front covers as described in Chapter 2, Section 2. Attach the multimeter between the black/red wire and the coil's mounting, to check the primary windings. No resistance figures are given by the manufacturer, but there should be continuity shown when the multimeter is set on the ohms range. Check the resistance in the secondary coil windings by connecting the multimeter (set on the K ohm range) between the spark plug cap and the coil's mounting; the reading given should be approximately 14 – 17 K ohm.

3   This test is unfortunately inconclusive and will only serve to give an indication of the unit's condition. If the readings obtained differ greatly, the unit should be given a full check by an authorized Suzuki dealer who will have the necessary equipment to carry out an accurate test. This involves testing on a spark gap tester which tests the coil under simulated normal riding conditions.

7   Ignition timing: checking and adjustment

1   Provision for adjustment of the timing does exist, but it should be noted that since none of the ignition components are subject to wear, it is unlikely that the timing will ever require adjustment. Accordingly there is no need to check the timing during the course of routine maintenance; this will only be necessary if a fault occurs in the ignition system.

2   Start by removing the crankcase left-hand cover and identifying the timing marks. These take the form of a scribed line on the rotor periphery and an index mark on the casing. If the marks are indistinct it is recommended to highlight them with a trace of white paint prior to the timing check. The timing should be checked using a stroboscope, preferably of the xenon type. Connect the stroboscope according to the manufacturer's instructions, start the engine and point the lamp at the timing marks on the periphery of the flywheel rotor. Raise the engine

speed to 6000 rpm and observe the marks; if the timing is correct they should be in exact alignment. Do not run the engine at this speed for any longer than is necessary to read the timing marks, otherwise engine damage could result. When the check is complete stop the engine and disconnect the equipment.

3   If adjustment of the timing is required this is made by altering the stator plate position. Note that access to the stator plate retaining screws is gained by removing the rotor, a task which will necessitate the use of an extractor tool. Refer to Chapter 1, Section 7 for details of rotor removal.

4   With the rotor removed, adjustment of the timing can be made by slackening the three stator plate screws and rotating the plate slightly to bring the timing to the correct position. Note that any adjustment will require rechecking by refitting the rotor and performing the timing check as described above.

8   HT lead and suppressor cap: general

1   Erratic running faults and problems with the engine suddenly cutting out in wet weather can often be attributed to leakage from the

high tension lead and spark plug cap. If this fault is present, it will often be possible to see tiny sparks around the lead and cap at night. One cause of this problem is the accumulation of mud and road grime around the lead, and the first thing to check is that the lead and cap are clean. It is often possible to cure the problem by cleaning the components and sealing them with an aerosol ignition sealer, which will leave an insulating coating on both components.

2 Water dispersant sprays are also highly recommended where the system has become swamped with water. Both these products are easily obtainable at most garages and accessory shops. Occasionally, the suppressor cap or the lead itself may break down internally. If this is suspected, the components should be renewed.

3 As the HT lead is permanently attached to the ignition coil, it is recommended that the renewal of the HT lead is entrusted to an auto-electrician who will have the expertise to solder on a new lead without damaging the coil windings.

9 Spark plug: general

Refer to Routine maintenance for information regarding examination, cleaning and setting the spark plug gap.

Chapter 4 Frame and forks

Contents

Specifications

Frame

Type	Welded tubular steel

Front forks

Type	Telescopic, coil sprung, hydraulically damped
Wheel travel..	200 mm (7.9 in)
Fork spring minimum free length	555 mm (21.85 in)
Fork oil level	207 mm (8.2 in)
Fork oil quantity – per leg	185 cc (6.5 fl oz)
Recommended fork oil..	SAE 10 fork oil

Rear suspension

Type	Suzuki Full Floater Suspension System
Swinging arm pivot bolt maximum runout...............................	0.6 mm (0.02 in)
Cushion lever bush ID service limit	20.3 mm (0.80 in)

Torque wrench settings

Component	kgf m	lbf ft
Handlebar clamp bolts	1.2 – 2.0	9 – 15
Steering stem bolt	3.5 – 5.5	25 – 40
Steering head bearing adjuster nut preload – slacken by $\frac{1}{4}$ – $\frac{1}{2}$ turn ...	4.0 – 5.0	29 – 36
Top and bottom yoke pinch bolts	2.0 – 3.0	15 – 22
Damper rod bolt...	1.5 – 2.5	11 – 18
Right-hand footrest bolts..	2.7 – 4.3	20 – 31
Swinging arm pivot bolt nut	5.0 – 8.0	36 – 58
Rear suspension unit nuts	4.0 – 6.0	29 – 43
Suspension linkage pivot bolt nuts	4.0 – 6.0	29 – 43
Brake torque arm nuts	1.0 – 1.5	7 – 11

1 General description

The duplex cradle frame is made of welded tubular steel and has a steel box-section swinging arm.

The front forks are of the conventional coil sprung, hydraulically damped type.

Rear suspension is by Suzuki's Full Floater type in which the suspension unit is not attached directly to the frame, but is connected by a 'floating' linkage. With this type of suspension the spring rate becomes progressively greater in response to increased wheel travel. The overall effect is that the suspension unit deflects a small amount in proportion to a large movement at the rear wheel.

2 Front forks: removal and refitting

1   Place a stout wooden box or blocks under the engine/gearbox unit so that the front wheel is raised clear of the ground. Remove the front wheel as described in Chapter 5, Section 2.

2   Unscrew the front brake cable retaining clamp from the top of the

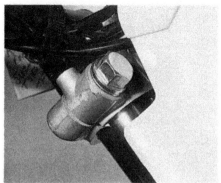

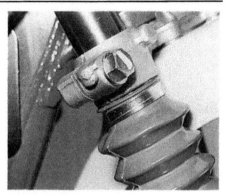

2.2 Slacken the headlamp cowling lower clamps ...

2.3a ... the top yoke pinch bolts ...

2.3b ... and bottom yoke pinch bolts to remove fork legs

fork lower leg and remove the speedometer drive cable from its guide. Slacken the headlamp cowling lower clamps.

3   The fork legs are retained by pinch bolts in the top and bottom yokes. Slacken the pinch bolts and slide the fork legs down and out of the yokes.

4   If a fork leg proves to be stuck, slacken the gaiter top clamp and slide the gaiter down exposing the fork stanchion. Apply penetrating oil to the yokes, hold the stanchion firmly and attempt to rotate it by hand to free the leg. It may be necessary to remove the pinch bolts completely and to spring the clamps apart slightly using a large flat-bladed screwdriver. Great care must be taken not to distort or break the clamp as this will necessitate the renewal of the yoke. If the leg is still reluctant to move, push a metal bar of suitable diameter through the spindle lug in the fork lower leg and tap downward on the protruding end of the bar to drive the fork leg from the yokes.

5   Once the legs have been removed, put them to one side to await further examination and dismantling as described in the following section.

6   To refit the fork legs, push them up through the yokes and head-lamp cowling ties until the top of the chrome stanchion is flush with the upper surface of the top yoke. It may be found necessary to remove the caps from the top of the stanchions and apply a small amount of grease or oil around the upper stanchion area to ease refitting. Refit the yoke clamp bolts together with any washers and cable guides fitted. Secure these bolts only lightly at this stage.

7   Refit the wheel as described in Chapter 5, Section 3, but only lightly tighten the spindle nut. Reconnect the brake and speedometer cables routing them through the appropriate cable guides.

8   Remove the machine supports and apply the rear brake hard whilst pushing down on the handlebars several times to settle the suspension components.

9   Tighten all nuts and bolts to their correct torque settings, starting at the wheel spindle and working up to the top yoke pinch bolts. This will ensure that the fork components can operate freely with no undue strain from an overtightened bolt or an awkwardly positioned component.

10   Check that the top of the fork stanchions have remained flush with the upper surface of the top yoke and refit the caps. Slide the gaiters up the stanchions so that they butt against the bottom yoke, then tighten the clamps.

11   Check the front fork operation, front brake adjustment and that all nuts and bolts are securely fastened before taking the machine on the road.

3   Front forks: dismantling and reassembly

1   Dismantle each leg separately so that there is no risk of interchanging components, and thus promoting undue wear. Slacken its upper and lower clamps and pull the gaiter off the fork leg.

2   Clamp the fork lower leg in a suitably padded vice to avoid marking the alloy of the leg. It is recommended that the fork leg is clamped at the wheel spindle boss to avoid the possibility of distorting the leg.

3   Remove the rubber cap from the top of the stanchion to expose the

metal top plug. Using a metal rod of suitable diameter, push down on the metal plug, against spring pressure, to expose the circlip. Carefully lever the circlip from its groove using a sharp instrument such as a small screwdriver and then gradually relax the downward pressure on the top plug, allowing the spring to push the top plug up and out of the stanchion. Great care must be exercised to prevent the top plug flying out under spring pressure, causing possible injury or damage; the aid of an assistant is strongly recommended.

4   Once the spring tension is released, remove the plug, spacer and washer, and withdraw the spring from the stanchion.

5   Drain the oil from the leg by inverting it and pumping the stanchion up and down several times to ensure that the maximum amount of oil is removed.

6   Reclamp the fork lower leg in the padded vice as described above. Use an allen key to unscrew the damper rod retaining bolt which is set in the base of the fork lower leg. In some cases the bolt will unscrew with ease, but it is quite likely that it will free itself from the lower leg and then rotate with the damper rod assembly. In such a case, obtain a length of wooden dowel, grind a coarse taper on one end which will engage in the head of the damper rod, and insert the dowel down into the bore of the stanchion. The services of an assistant will now be required. Clamp a self-locking wrench to the protruding end of the dowel and with the assistant preventing the dowel from turning and simultaneously applying pressure via the dowel to the head of the damper rod, the damper rod will be locked in place so that the retaining bolt can be unscrewed. When working alone, use a longer length of dowel which can be clamped in the vice. By pushing down on the fork lower leg with one hand it should be possible to lock the damper rod firmly enough for the retaining bolt to be unscrewed.

7   With the damper rod bolt removed, carefully pull the stanchion out of the lower leg. Invert the stanchion to tip out the damper rod and spring, then invert the lower leg to tip out the damper rod seat.

8   The fork oil seal should be renewed whenever the stanchion is removed and must be renewed if it is disturbed as the means used for removing the seal will almost certainly damage it. The seal is retained by a circlip which must be removed using a small, flat-bladed screwdriver to ease it away from its groove in the fork lower leg. Use a large, flat-bladed screwdriver to lever the seal from its housing. Take care not to scratch the internal surface of the seal housing with the edge of the screwdriver blade, and do not apply excessive pressure or there is a risk of the upper edge of the fork lower leg being cracked or distorted. If the seal appears difficult to move, heat the leg by pouring boiling water over its outer surface. This will cause the alloy leg to expand sufficiently to loosen the seal. Do not forget to displace the spacer fitted underneath the oil seal.

9   Reassembly is a reversal of the dismantling procedure, but careful attention must be paid to certain points.

10   Insert the spacer into the top of the lower leg so that its raised edge faces upwards. Tap the spacer down until it rests in the machined shoulder. Clamp the lower leg securely in a vice by means of the wheel spindle boss. Coat the inner and outer diameters of the seal with the recommended fork oil and push the seal squarely into the bore of the fork lower leg by hand. Ensure that the seal is fitted squarely, then tap it fully into position using a hammer and a suitably sized drift such as a socket spanner, as shown in the accompanying photograph. The drift should bear only on the hard outer diameter of the seal, never on the sealing lips themselves, and should have a smooth undamaged surface

3.8a Carefully lever the circlip out of the lower leg

3.8b Use a flat-bladed screwdriver to lever the oil seal out of the lower leg

3.10a Insert the spacer into the lower leg, raised edge upwards

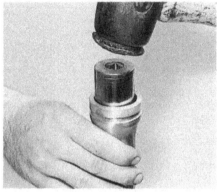

3.10b Tap the new oil seal into its housing, then secure with the circlip

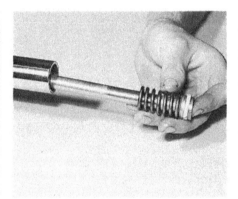

3.11a Refit the damper rod and spring into the stanchion ...

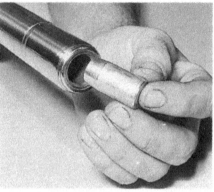

3.11b ... then refit the damper rod seat to its protruding end

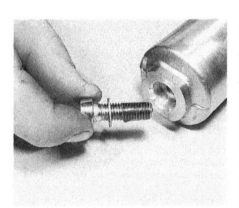

3.12 Secure the assembly with the damper rod bolt, using a locking compound on its threads

3.15 Measure oil level with dipstick cut to correct length

3.16a Refit the fork spring (closer coils upwards) ...

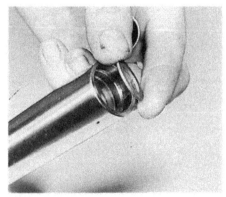

3.16b ... the washer ...

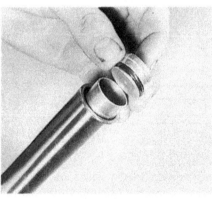

3.16c ... spacer and top plug

3.16d Ensure that the circlip is seated correctly in its groove ...

where it comes into contact with the seal. Tap the seal into the bore of the lower leg just enough to expose the circlip groove. Refit the retaining circlip securely in its groove.

11   Slide the damper rod rebound spring into place under the head of the damper rod and insert the damper rod assembly into the stanchion. Push the damper rod down the length of the stanchion until its end projects from the stanchion lower end. Refit the damper rod seat over the end of the damper rod, using a smear of grease to stick it in place.

12   Smear the sliding surface of the stanchion with a light coating of fork oil and carefully insert the stanchion into the lower leg, taking great care not to damage the sealing lips of the oil seal. Push a fork spring or the length of dowel used on dismantling down into the stanchion and apply pressure on this to ensure that the damper rod or its seat is pressed firmly into the base of the lower leg. Check that the threads of the damper rod bolt are clean and dry, apply a few drops of thread-locking compound and fit the damper rod bolt. Do not forget the sealing washer fitted under the head of the bolt. Tighten the bolt only partially at first, using an Allen key of suitable size. Maintain pressure on the head of the damper rod and push the stanchion firmly as far down into the lower leg as possible to centralise the damper rod in the stanchion. The damper rod bolt can then be tightened firmly to the recommended torque setting. Withdraw the spring or dowel from the stanchion.

13   Smear a light coating of grease over the stanchion sliding surface to provide additional protection for the stanchion and seal against moisture or dirt. Fit the gaiter and its retaining clamps but tighten only the bottom clamp at this stage. The upper clamp should only be tightened when the fork leg has been fitted in the yokes and correctly positioned.

14   Fill the leg with the correct amount of the specified grade of fork oil (see Specifications) and ensure that exactly the same amount is put into each leg. Note that a fork oil level is also given.

15   To measure the oil level in each leg, make up a dipstick from a piece of wire cut to the required length. Check that the fork spring has been removed and push the stanchion fully into the fork lower leg. Pour the specified amount of oil into the fork leg and allow the level to settle. Add or remove oil as necessary. Once the correct oil level has been established in this way, the fork legs can be refilled to that level during routine maintenance when the forks are not completely dismantled. This will automatically allow for the presence of any residual fork oil which cannot be removed without fully dismantling the fork leg and will ensure smooth and consistent fork action.

16   When the fork oil has been poured in, and the level checked, pull the stanchion out of the leg as far as possible and insert the fork spring noting that the closer spaced coils must be at the top. To refit the fork top plug, enlist the services of an assistant to ease the operation, which is a straightforward reversal of the dismantling procedure. Note that the manufacturer recommends that the O-ring be renewed during reassembly. Take great care not to damage the sealing O-ring when fitting to the top plug and ensure that the retaining circlip is securely engaged in its locating groove in the stanchion.

17   Refit the fork leg into the fork yokes following instructions given in Section 2 of this Chapter.

---

4   Front forks: examination and renovation

1   Carefully clean and dry all the components of the fork leg. Lay them out on a clean work surface and inspect each one, looking for excessive wear, cracks, or other damage. All traces of oil, dirt, and swarf should be removed, and any damaged or worn components renewed.

2   Examine the sliding surface of the stanchion and the internal surface of the lower leg, looking for signs of scuffing which will indicate that excessive wear has taken place. Slide the stanchion into the lower leg so that it seats fully. Any wear present will be easily found by attempting to move the stanchion backwards and forwards, and from side to side, in the bore of the lower leg. It is inevitable that a certain degree of slackness will be found, especially when the test is repeated at different points as the stanchion is gradually withdrawn from the lower leg, and it is largely a matter of experience to assess accurately the amount of wear necessary to justify renewal of either the stanchion or the lower leg. It is recommended that the two components be taken to a motorcycle dealer for an expert opinion to be given if there is any doubt about the degree of wear found. Note that while wear will only become a serious problem after a high mileage has been covered, it is essential that such wear is rectified by the renewal of the components concerned if the handling and stability of the machine are not to be impaired.

3   Check the outer surface of the stanchion for scratches or roughness; it is only too easy to damage the oil seal during the reassembly if these high spots are not eased down. The stanchions are unlikely to bend unless the machine is damaged in an accident. Any significant bend will be detected by eye, but if there is any doubt about straightness, roll down the stanchion tubes on a flat surface such as a sheet of plate glass. If the stanchions are bent, they must be renewed. Unless specialised repair equipment is available it is rarely practicable to effect a satisfactory repair to a damaged stanchion.

4   Check the stanchion sliding surface for pits caused by corrosion; if the gaiters are intact such damage will be rare. Such pits should be smoothed down with fine emery paper and filled, if necessary, with Araldite. Once the Araldite has set fully hard, use a fine file or emery paper to rub it down so that the original contour of the stanchion is restored.

5   After an extended period of service, the fork springs may take a permanent set. If the spring lengths are suspect, then they should be measured and the readings obtained compared with the service limits given in the Specifications Section of this Chapter. It is always advisable to fit new fork springs where the length of the original items has decreased beyond the service limit given. Always renew the springs as a set, never separately.

3.16e ... then fit the cap

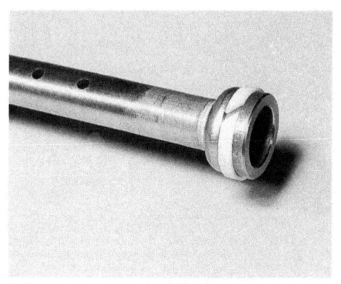

4.6 Check condition of piston ring fitted to damper rod

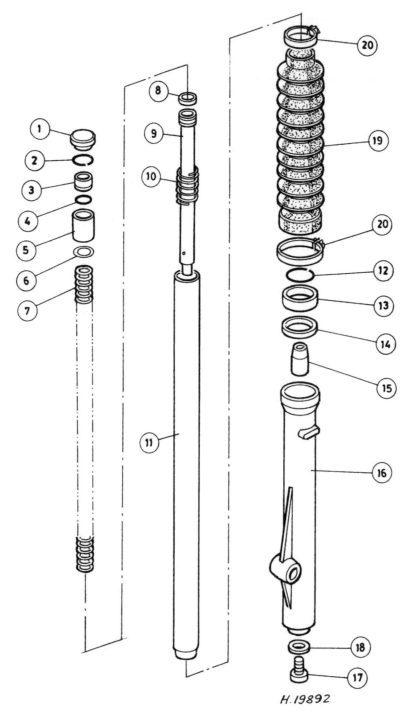

H.19892

Fig. 4.1 Front forks

1	Cap	6	Spring seat	11	Stanchion	16	Lower leg
2	Circlip	7	Spring	12	Circlip	17	Damper rod bolt
3	Top plug	8	Piston ring	13	Oil seal	18	Sealing washer
4	O-ring	9	Damper rod	14	Spacer	19	Gaiter
5	Spacer	10	Rebound spring	15	Damper rod seat	20	Gaiter clamp – 2 off

6   The piston ring fitted to the damper rod may wear if oil changes at the specified intervals are neglected. If damping has become weakened and does not improve as a result of an oil change, the piston ring should be renewed. Check also that the oilways in the damper rod have not become obstructed.

7   Closely examine the gaiter for splits or signs of deterioration. If found to be defective, it must be renewed as any ingress of dirt will rapidly accelerate wear of the oil seal and fork stanchion. It is advisable to renew any sealing washers fitted beneath bolt heads as a matter of course. The same applies to the O-rings fitted to the fork top plugs.

---

5   Steering head assembly: removal and refitting

1   The steering head will rarely require attention unless it becomes necessary to renew the bearings or if accident damage has occurred.

The full dismantling sequence is described here, but it is quite in order to avoid as much dismantling as possible by careful manoeuvring of the ancillary components. Much depends on a common sense approach and ingenuity on the part of the owner.

2   Remove the front wheel and front forks as described in Chapter 5 and Section 2 of this Chapter respectively. The front mudguard is retained by four bolts located underneath the bottom yoke. Remove the bolts and withdraw the mudguard making a note of the spacer positions as a guide to refitting.

3   To prevent short circuits, disconnect the battery (negative lead first) before removing any electrical leads.

4   The headlamp cowling is retained at the top by two screws which thread into extensions of the instrument console mounting bracket and at the bottom by two rings which wrap around the fork stanchions and lugs on the cowling base.

5   Release the cowling and ease it forward to give access to the electrical components behind it. Disconnect the headlamp and parking lamp bulbs and remove the headlamp and cowling as a complete unit.

6   Remove the fuel tank as described in Chapter 2 or protect it with an old blanket or similar padding to prevent damage to the paintwork. The handlebars are secured to the top yoke by two split clamps. Slacken and remove the four bolts, withdraw the upper clamp halves and carefully move the handlebars backwards, clear of the steering head area, without placing undue strain on the cables or wiring. Withdraw the front brake cable from its guides and tape it to a convenient part of the frame, out of harm's way. The ignition switch assembly is retained by the handlebar bolts and can be removed once its wiring connector has been separated.

7   The speedometer and tachometer drive cables are retained by knurled rings, located underneath the instrument console. Unscrew both rings and pull both cables clear of the instruments and steering head area.

8   The instrument console, complete with mounting bracket, can now be removed by disconnecting the wiring connectors from the main loom and removing the two retaining bolts in the top yoke.

9   Remove the large chromium plated bolt from the steering head. Using a soft-faced mallet, give the top yoke a gentle tap to free it from the steering head and lift it clear of the steering stem.

10   Support the weight of the bottom yoke and, using a C-spanner of the correct size, remove the steering head bearing adjusting ring. If a C-spanner is not available, a soft metal drift may be used in conjunction with a hammer to slacken the ring.

11   Remove the dust excluder and the cone of the upper bearing. The bottom yoke, complete with steering stem, can now be lowered from position. Ensure that any balls that fall from the bearings as the bearing races separate are caught and retained. It is quite likely that only the balls from the lower bearing will drop free, since those of the upper bearing will remain seated in the bearing cup. Refer to the following section for details of bearing examination.

12   Note that on refitting it is advisable to position all eighteen balls of

5.2 Remove four bolts to release mudguard from bottom yoke

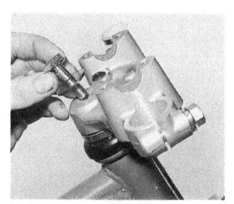

5.9a Remove the steering head bolt ...

5.9b ... and lift off the top yoke

5.10 Support the bottom yoke and remove the adjuster nut ...

5.11a ... the dust excluder and bearing top cone ...

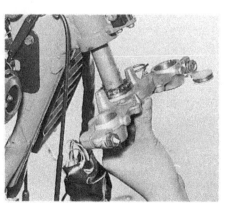

5.11b ... then lower the bottom yoke out of the steering head

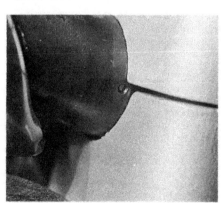

5.16 Align handlebar punch mark with rear split clamp on refitting

the lower bearing around the bearing cone before inserting the steering stem fully into the steering head. Retain these balls in position with grease of the recommended type and fill both bearing cups with the same type of grease.

13   With the bottom yoke pressed fully home into the steering head, place the twenty-two balls into the upper bearing cup and fit the bearing cone followed by the dust excluder. Refit the adjusting ring and tighten it, finger-tight. The ring should now be tightened firmly to seat the bearings; the manufacturer recommends a torque setting of 4.0 – 5.0 kgf m (29 – 36 lbf ft). Turn the bottom yoke from lock to lock five or six times to settle the balls, then slacken the adjusting ring by a quarter to a half turn. This will form a basis for adjustment when reassembly is complete, as described in Routine maintenance.

14   Reassembly is a straightforward reversal of the dismantling procedure but careful attention must be paid to the following points:

15   When refitting the top yoke onto the steering stem, it is advisable to temporarily refit the fork legs. This will ensure correct alignment of the upper and lower yokes before tightening the chromium plated steering stem bolt to the correct torque setting.

16   When refitting the handlebars align the punch mark on the handlebar with the rear join of the clamps. The securing bolts must be tightened evenly to the recommended torque setting, keeping an equidistant gap between the upper clamp halves and the top yoke.

17   Check the steering head bearing adjustment as described in Routine maintenance, after the steering head has been reassembled and the forks and front wheel refitted.

18   Finally, whilst refitting and reconnecting all disturbed components, take care to ensure that all control cables, drive cables, electrical leads, etc are correctly routed and that proper reference is made to the list of torque wrench settings given in the Specifications Section of this Chapter and of Chapter 5. Check that the headlamp beam height has not been disturbed and ensure that all controls and instruments function correctly before taking the machine on the public highway.

6   Steering head bearings: examination and renovation

1   Before commencing reassembly of the steering head component parts, take care to examine each of the steering head bearing races. The ball bearing tracks of their respective cup and cone bearings should be polished and free from any indentations or cracks. If wear or damage is evident, the cups and cones must be renewed as complete sets.

2   Carefully clean and examine the balls contained in each bearing assembly. These should also be polished and show no signs of surface cracks or blemishes. If any one ball is found to be defective, the complete set should be renewed. Remember that a complete set of these balls is relatively cheap and it is not worth the risk of refitting items that are in doubtful condition. Note that two different sizes of ball are used; the upper race being fitted with 6 mm ($\frac{1}{4}$ in) balls and the lower race with 8 mm ($\frac{5}{16}$ in) items. Be careful at all times not to confuse the two sizes and ensure that each race is fitted with balls of the correct size.

3   Twenty-two balls are fitted in the top bearing race and eighteen in the lower. This arrangement will leave a gap between any two balls but an extra ball must not be fitted, otherwise the balls will press against each other, thereby accelerating wear and causing the steering action to be stiff.

4   The bearing cups are a drive fit in the steering head and may be removed by passing a long drift through the inner bore of the steering head and drifting out the defective item from the opposite end. The drift must be moved progressively around the cup to ensure that it leaves the steering head evenly and squarely.

5   The lower of the two cones fits over the steering stem and may be removed by carefully drifting it up the length of the stem with a flat-ended chisel or a similar tool. Again, take care to ensure that the cone is kept square to the stem.

6   Fitting of the new bearing cups is a straightforward procedure whilst taking note of the following points. Ensure that the cup locations within the steering head are clean and free of rust; the same applies to the steering stem. Lightly grease the stem and head locations to aid fitting of the cups or cone and drift each item into position whilst keeping it square to its location. Fitting of the cups into the steering head will be made easier if the opposite end of the head to which the race is being fitted has a wooden block placed against it to absorb some of the shock as the drift strikes the cup.

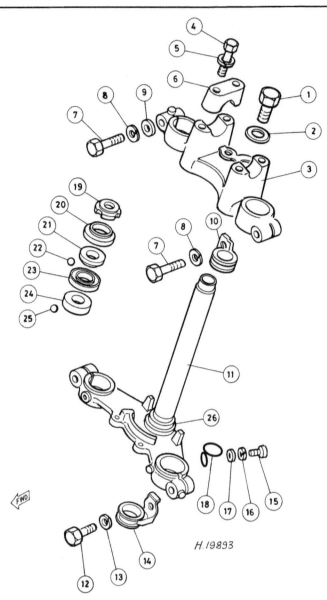

Fig. 4.2 Steering head assembly

1	Bolt	14	Cable guide
2	Washer	15	Screw
3	Top yoke	16	Spring washer
4	Bolt – 4 off	17	Washer
5	Spring washer – 4 off	18	Cable guide
6	Handlebar clamp – 2 off	19	Adjuster nut
7	Bolt – 2 off	20	Dust excluder
8	Spring washer – 2 off	21	Top cone
9	Washer	22	Top bearing balls – 22 off
10	Cable guide	23	Top cup
11	Steering stem/lower yoke	24	Bottom cup
12	Bolt – 2 off	25	Bottom bearing balls – 18 off
13	Spring washer – 2 off	26	Bottom cone

7   Frame: examination and renovation

1   The frame is unlikely to require attention unless accident damage has occurred. In some cases, renewal of the frame is the only satisfactory remedy if it is badly out of alignment. Only a few frame specialists have the jigs and mandrels necessary for resetting the frame to the required standard of accuracy, and even then there is no easy means of assessing to what extent the frame may have been over-stressed.

2   After the machine has covered a considerable mileage, it is advisable to examine the frame closely for signs of cracking or splitting at the

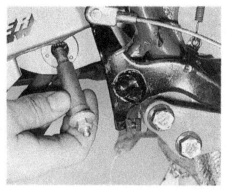  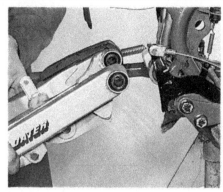

8.3 Remove the cushion rod lower mounting nut and pivot bolt

8.4a Remove nut and withdraw pivot bolt ...

8.4b ... to permit removal of the swinging arm

welded joints. Rust corrosion can also cause weakness at these joints. Minor damage can be repaired by welding or brazing, depending on the extent and nature of the damage.

3 Remember that a frame which is out of alignment will cause handling problems and may even promote 'speed wobbles'. If misalignment is suspected, as a result of an accident, it will be necessary to strip the machine completely so that the frame can be checked, and if necessary, renewed.

8 Swinging arm: removal and refitting

1 Depending on the work being undertaken, note that the swinging arm and suspension linkage can be removed as a complete unit rather than as separate assemblies. The removal procedure is greatly reduced if only the swinging arm is to be removed.

2 Remove the rear wheel following the instructions given in Chapter 5, Section 3.

3 Remove the cushion rod lower mounting nut and withdraw the bolt enabling the swinging arm to hang down so that the cushion rod is clear of its mounting.

4 Slacken and remove the swinging arm pivot nut, complete with its washer. Withdraw the swinging arm pivot bolt and manoeuvre the swinging arm rearwards away from the frame mounting points and drive chain. If the bolt proves stubborn, apply a liberal quantity of penetrating fluid whilst rotating the bolt. Allow a short time for the fluid to work then displace the bolt using a hammer and soft metal drift, taking care not to damage the bolt threads or bushes.

5 Examine the swinging arm pivot bolt, the lower cushion rod bolt and the passages through which they both run. Ensure that these are completely free from dirt and corrosion, removing any traces of old hardened grease with a high flash-point solvent.

6 Check the pivot bolt shaft, swinging arm bushes and dust seals as described in the following section and renew any worn items. Examine the lower cushion rod bolt and bearing following the instructions given in Section 10 of this Chapter and regrease accordingly. Smear a liberal quantity of high melting-point grease along the length of the pivot bolt shank and its passage to keep corrosion at bay and aid any future removal of the swinging arm or the engine/gearbox unit.

7 On refitting the swinging arm, ensure that the sleeves and dust seals are in place and that the rubber chain buffer is located in its correct position. Offer up the swinging arm to the frame and retain it by pushing through the pivot bolt from left to right. Refit the washer and retaining nut.

8 Swing the assembly upwards until the cushion rod eye aligns with the holes in the swinging arm, whilst making sure that the headed spacers fitted on each side of the mounting eye have not been displaced. Insert the pivot bolt from right to left and secure with the washer and nut.

9 Tighten both nuts to the correct torque setting given in the Specifications section of this Chapter.

10 Check that the drive chain routing is correct and refit the rear wheel to the machine as described in Chapter 5. Always ensure that the rear brake, stop lamp switch and chain tension are correctly adjusted and working properly before using the machine on the road; refer to Routine maintenance for details.

9 Swinging arm: examination and renovation

1 Withdraw the swinging arm from the frame as described in the previous section and remove all traces of dirt and grease from its components using a high flash-point solvent.

2 Inspect the structure of the swinging arm for signs of distortion, cracks or any other damage which may lead to component failure. Pay particular attention to the welded areas between component parts. If fracture or distortion is suspected, return the swinging arm to an authorized Suzuki dealer who will confirm whether renewal is necessary.

3 Remove the chain buffer from the left-hand pivot lug and inspect it for signs of excessive wear. If it no longer serves as an adequate means of protecting the swinging arm against interference from the drive chain or if it is showing signs of cracking or splitting, it must be renewed.

4 The central metal sleeves of the pivot bushes should push out easily. If they prove stubborn, very carefully tap them out using a hammer and suitably sized drift. If they are badly scored or corroded, they must be renewed.

5 Carefully prise out the outer dust seals with a flat-bladed screwdriver. The inner ones can either be removed similarly or extracted along with the bushes. Examine the dust seals and renew if at all worn or damaged; ideally these seals should be renewed whenever they are disturbed.

6 The bushes should not be removed unless they are worn, as the removal process is almost certain to damage them. It follows that new bushes should be fitted whenever the existing ones are disturbed. Remove the bushes using a hammer and suitably sized drift, forcing the bush out of the pivot lug. If this fails, a drawbolt tool can be applied as shown in the accompanying illustration.

7 The fitting of new bushes is a reversal of the removal procedure.

9.4 Swinging arm inner sleeves are a push fit in their bushes

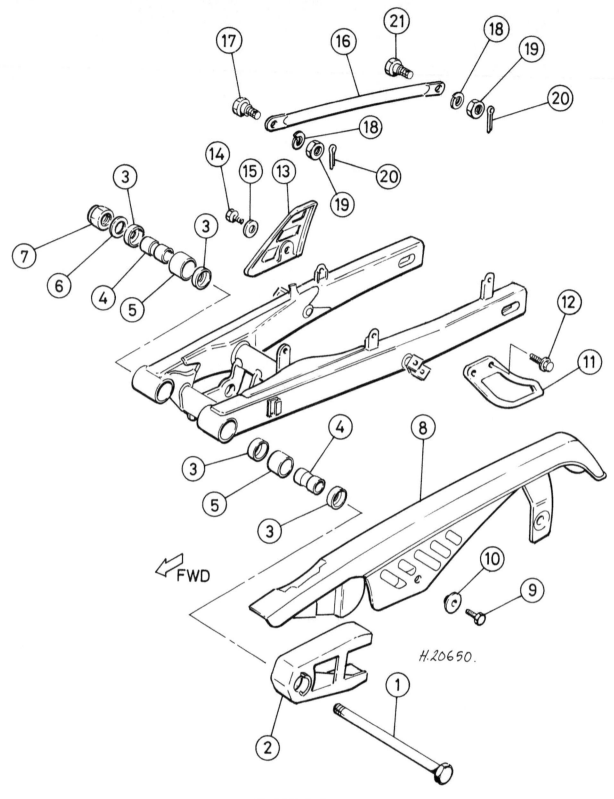

Fig. 4.3 Swinging arm

1 Pivot bolt	7 Nut	13 Foot guard	19 Nut – 2 off
2 Chain buffer	8 Chainguard	14 Bolt	20 Split pin – 2 off
3 Dust seal – 4 off	9 Bolt – 3 off	15 Washer	21 Bolt
4 Sleeve – 2 off	10 Washer – 3 off	16 Brake torque arm	
5 Bush – 2 off	11 Chain guide	17 Bolt	
6 Washer	12 Bolt – 2 off	18 Spring washer – 2 off	

H.20650.

FWD

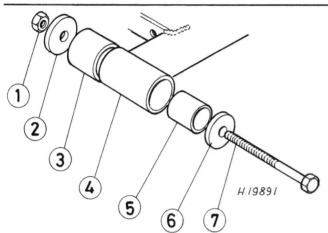

Fig. 4.4 A drawbolt tool can be used to insert or remove swinging arm bushes

1	Nut	5	Bush
2	Washer	6	Washer
3	Tubing	7	Bolt
4	Swinging arm		

Thoroughly clean the bush housing in the swinging arm and remove any traces of corrosion or burrs to the metal. Coat the housing with a smear of grease to aid refitting. Lightly grease and insert the new bush using either a drawbolt tool or by tapping it into place with a hammer and tubular drift which has an external diameter slightly less than that of the bush. Draw or tap the bush into position ensuring that it is absolutely square to the housing at all times.

8 Pack the dust seals and central spacers with high melting-point grease before installing them. This will not only help smooth the pivoting action of the swinging arm but will prevent any foreign matter such as dirt or dust thrown up by the rear wheel from entering the pivot assembly.

9 Check the swinging arm pivot bolt for wear. If the shank of the bolt is badly scored or stepped it must be renewed. The straightness of the bolt can be determined by placing it on V-blocks and measuring the runout on its shank using a dial gauge. Alternatively, rolling the bolt along a flat surface such as a sheet of plate glass whilst attempting to insert a feeler gauge of 0.6 mm (0.02 in) beneath it will serve as an adequate means of assessing whether or not the bolt is bent. Renew the bolt if the amount of runout exceeds the service limit. A bent pivot bolt will prevent the swinging arm from moving smoothly about its axis.

10 Finally, remember to refit the chain buffer before the swinging arm is refitted to the machine.

10 Suspension linkage: removal, examination, renovation and refitting

1 Remove the swinging arm following the instructions given in Section 8.

2 Remove the dualseat and sidepanels as described later in this Chapter to gain access to the suspension unit upper mounting. The mounting bolt which passes through a rubber bush in the top suspension eye also carries a mounting for the oil tank. Remove the nut and lever the oil tank extension to one side, taking care not to break or damage the plastic. Unscrew the second (larger) nut and withdraw the bolt from the frame.

3 Slacken and remove the two nuts securing the cushion lever brackets to the frame and withdraw the bolts. The brackets are mounted on extension lugs, welded to the frame at the rear of the engine/gearbox unit. With the securing bolts thus removed, manoeuvre the suspension linkage and suspension unit rearwards, and withdraw it from the frame. Place the assembly on the bench to await further dismantling and examination.

4 The individual components can be separated as follows: Remove the suspension unit lower mounting nut and bolt to free the unit from the cushion lever; note the presence of the two headed spacers. The cushion lever brackets can be disconnected from the cushion lever after the nut and pivot bolt have been removed, noting the washers placed each side of the cushion lever. This leaves the cushion lever attached to the cushion lever rod. The two components can be separated after the nut and pivot bolt have been removed; note the presence of the headed spacers. Thoroughly clean each part, removing all traces of dirt, corrosion or old grease and inspect all components for signs of wear. Cracks in the casting or any damage which will impair the effectiveness of the component will indicate the need for renewal.

5 Remove the inner sleeve of the cushion lever bush and measure the inside diameter of the bush itself. If the bush exceeds the service limit at any point it must be renewed, together with the inner sleeve. Use a similar method to that described for removing and refitting the swinging arm bushes in Section 9 of this Chapter. Note that the dust seals fitted each side of the bush should be levered from position using a flat-bladed screwdriver prior to bush removal. Once the bush has been removed check the operation of the grease nipple by connecting a grease gun to the nipple and forcing grease through into the bearing housing. If the nipple shows signs of blockage it should be unscrewed from the casting and all traces of hardened grease or dirt removed.

6 The bearings fitted to the remaining bores of the cushion rod and the cushion rod lever are of the spherical type and are used to ensure correct alignment between the various components. Check each bearing by feel; they should rotate freely with no sign of notchiness. If removal is required lever out the dust seals and use a socket spanner of similar diameter to the bearing outer race to drive the bearing squarely out of its housing.

7 If new bearings are being fitted these should be driven or pressed into place and packed with fresh grease. Renew any dust seals disturbed during removal or which show signs of damage. Note that failure to renew a worn or damaged seal will allow dirt to enter the pivot assembly.

8 Carefully inspect the linkage pivot bolts for wear on their shanks. It will be necessary to renew the bolts if deep scoring is present. The two bolts with grease nipples inserted into their heads have a small hole along their shank to allow the lubricant to enter the bearing without any dismantling being necessary. Ensure that the hole in each pivot bolt is not blocked and will allow the grease to pass through.

9 The suspension unit comprises an hydraulic damper, coil spring and a rubber bush, forming its top mounting. It is not necessary to dismantle

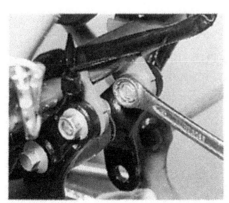

10.3 Cushion lever brackets are secured by a nut and bolt to the frame

10.4 Suspension linkage can be removed as a single unit and dismantled for attention to its pivot bearings

10.5 Cushion lever bush inside diameter measurement

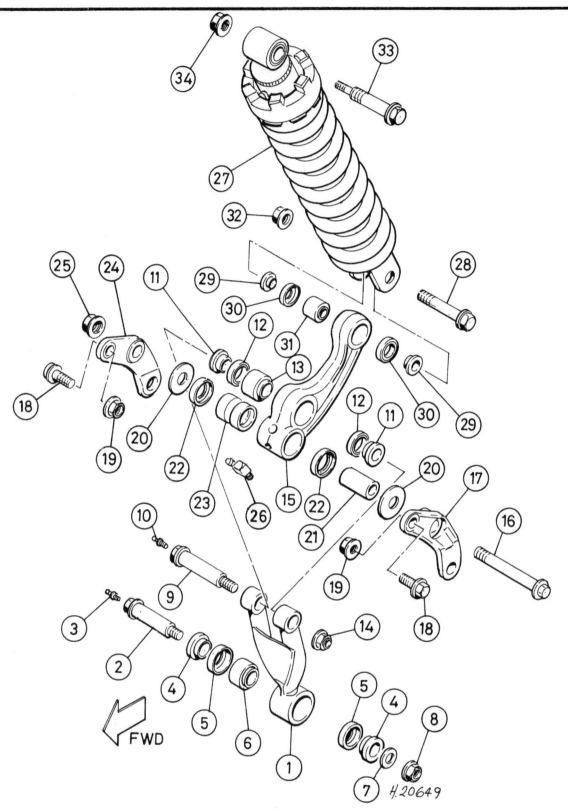

Fig. 4.5 Rear suspension linkage

1 Cushion lever rod	10 Grease nipple	18 Bolt – 2 off	26 Grease nipple
2 Pivot bolt	11 Headed spacer – 2 off	19 Nut – 2 off	27 Suspension unit
3 Grease nipple	12 Dust seal – 2 off	20 Washer – 2 off	28 Pivot bolt
4 Headed spacer – 2 off	13 Spherical bearing	21 Sleeve	29 Headed spacer – 2 off
5 Dust seal – 2 off	14 Nut	22 Dust seal – 2 off	30 Dust seal – 2 off
6 Spherical bearing	15 Cushion lever	23 Bush	31 Spherical bearing
7 Washer	16 Pivot bolt	24 Right-hand cushion	32 Nut
8 Nut	17 Left-hand cushion	lever bracket	33 Pivot bolt
9 Pivot bolt	lever bracket	25 Nut	34 Nut

or remove the complete suspension linkage if only the suspension unit is to be removed. After removal of the rear wheel, seat and sidepanels, withdraw the two suspension unit securing bolts (having disconnected the oil tank front mounting) and carefully manoeuvre the unit rearwards past the linkages and swinging arm.

10   Replacement parts for the rear suspension unit are not available. If the damper fails or starts to leak, the complete suspension unit must be renewed. This applies equally if the damper rod has become bent or if the rubber bush located in the top mounting eye has worn.

11   The shock absorber unit has an adjustable preload setting which can be altered to compensate for varying loads. This setting can be easily altered without removing the unit from the machine. Use a suitably sized C-spanner to slacken the lock ring. The adjuster ring (located immediately below the lock ring) can be turned clockwise to increase the spring tension and stiffen the rear suspension, or anticlockwise to lessen the spring tension and therefore soften the ride. Remember to tighten the lock ring after any adjustments have been made.

12   Reassemble the linkage components and refit the suspension linkage and suspension unit to the machine by a reversal of the removal procedure. Tighten the linkage bolts and pivot shaft nuts to the correct torque settings given in the Specifications Section of this Chapter.

13   Refit the swinging arm, seat and sidepanels following the instructions given in this Chapter and refit the rear wheel as described in Chapter 5.

14   Finally apply grease to the suspension linkage bearings via the grease nipples provided.

11 Prop stand: general

1   The prop stand is bolted to a welded lug on the frame. A return spring is fitted to ensure the stand retracts when the weight of the machine is taken off the stand. When fully retracted the stand should be tight against its stop and well out of the way.

2   Check that the nut retaining the pivot bolt is fully tightened and that the pivot is well lubricated. The return spring should be free from fatigue and in good condition. Remember that a serious accident is almost inevitable if the stand extends whilst the machine is in motion.

12 Footrests: general

1   The right-hand rider's footrest is mounted on a separate bracket, retained by two bolts to the frame, and the left-hand footrest is mounted directly on the frame. The footrests are hinged on clevis pins enabling them to fold upwards in the event of the machine being accidentally dropped.

2   Should removal of a footrest be necessary, remove the split pin and clevis pin paying attention to the way in which the return spring is fitted.

3   If accident damage has occurred, check that the damaged component is not cracked or broken. Such damage may be repaired by welding but since this will destroy the finish, renewal is usually the most satisfactory course of action. If a footrest is merely bent, it can be straightened after the affected area has been heated to a cherry red colour using a blowlamp or welding torch. Again the painted surface will be destroyed but can be repainted quite easily.

4   The pillion footrests pivot on clevis pins secured by split pins. A damaged footrest can be repaired as described above, remembering to first remove the footrest rubber prior to the application of heat.

5   Note that if there is any evidence of fatigue before or after straightening, it is advisable that the damaged component is renewed. If a footrest breaks, loss of machine control is almost inevitable.

6   When refitting the right-hand footrest bracket, ensure that both bolts are tightened to the correct torque setting. In all cases ensure that a new split pin is used to secure the clevis pin and that the legs of the pin are correctly spread.

7   A few drops of oil applied to the footrest pivots at regular intervals will ensure that corrosion does not set in, causing premature wear.

13 Rear brake pedal: general

1   To remove the brake pedal, first disconnect the rear stop lamp switch spring. Remove the split pin, clevis pin and washer which retain the brake rod to the pedal and separate the two components. Unhook the return spring and withdraw the split pin and washer from the brake pedal pivot. Remove the brake pedal from the frame, noting the location of the return spring.

2   The examination and renovation procedures are the same as those given for the footrests in the previous section.

3   Reassembly is a straightforward reversal of the removal procedure. Lubricate the brake pedal pivot with grease and fit new split pins to the pivot and rod connection. Ensure that the rear brake and stop lamp switch are functioning correctly before using the machine on the road. If necessary, adjust brake pedal height as described in Routine maintenance.

14 Steering lock: general

1   A steering lock is attached to a lug on the steering head by means of two screws, each with a spring washer beneath its head. When in the locked position, a bar extends from the body of the lock and abuts against a projection which forms part of the casting of the lower yoke. This effectively prevents the handlebars from being turned once they are set on full lock.

2   If the lock malfunctions it must be renewed; repairs are impracticable. When the lock has been renewed, a key for it should be carried when the machine is in use.

15 Instrument drive cables: examination

1   It is advisable to detach the speedometer and tachometer drive cables from time to time in order to check whether they are adequately lubricated and whether the outer cables are compressed or damaged at any other point along their run. A jerky or sluggish movement at the instrument head can often be attributed to a cable fault.

2   The cables are secured at both ends by large knurled rings which must be tightened or slackened using a suitable pair of pliers. Do not overtighten the knurled rings, or they will crack, necessitating renewal of the complete cable.

3   To grease the cable, uncouple both ends and withdraw the inner cable. After removing any old grease, clean the inner cable with solvent and examine the cable for broken strands or other damage. Do not

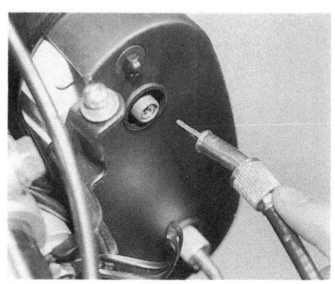

15.2 Instrument cables are retained by knurled rings

check the cable for broken strands by passing it through the fingers or palm of the hand, this may well cause a painful injury if a broken strand snags the skin. It is best to wrap a piece of rag around the cable and pull the cable through it, any broken strands will snag on the rag.

4   Regrease the cable with high melting-point grease, taking care not to grease the last six inches closest to the instrument head. If this precaution is not observed, grease will work into the instrument and immobilise the sensitive movement.

5   When refitting drive cables, always ensure that they have smooth, easy runs to minimise wear, and that the oil seal fitted to the inner cable's lower end is correctly seated. Check that the cables are secured where necessary by any clamps or ties provided to keep the cables away from any hot or moving parts.

16   Instrument heads: removal, examination and refitting

1   These instruments must be carefully handled at all times and must never be dropped or held upside down. Dirt, oil, grease and water all have an equally adverse effect on them, and so a clean working area must be provided if they are to be removed.

2   The instrument heads are very delicate and should not be dismantled at home. In the event of a fault developing, the instrument should be entrusted to a specialist repairer or a new unit fitted. If a replacement unit is required it is well worth trying to obtain a good secondhand item from a motorcycle breaker in view of the high cost of a new instrument.

3   Remember that a speedometer in correct working order is a statutory requirement in the UK. Apart from this legal necessity, reference to the odometer readings is the most satisfactory means of keeping pace with the maintenance schedules.

4   Remove the headlamp cowling and disconnect the drive cables by slackening their knurled retaining rings. Slacken and remove the two bolts securing the instrument panel to the top yoke, disconnect the supply wires and withdraw the panel assembly, complete with mounting bracket. Unscrew the two domed nuts and remove the panel base and mounting bracket. Separate the top cover from the housing by removing the two self-tapping screws from the base of the unit and withdrawing the tripmeter reset knob. Each instrument is secured to the housing by two screws with plain washers.

5   Refitting is the reverse of the above procedure.

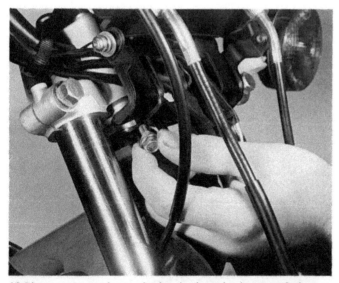

16.4 Instrument panel mounting bracket is retained to top yoke by two bolts

17   Instrument drives: removal, examination and refitting

1   The speedometer drive is located in the front brake backplate. It is therefore necessary to withdraw the front wheel from the machine in order to gain access to the drive components, see Chapter 5, Section 2.

2   To dismantle the drive assembly, lift out the drive gear from the inner face of the backplate. Turn the backplate over and remove the small grub screw from the boss which houses the driven gear components. Pull the outer bush from its housing and remove the speedometer driven pinion, taking care not to lose the two thrust washers which fit over each end.

3   Carefully clean and inspect all the component parts. Any wear or damage will be immediately obvious and should be rectified by the renewal of the part concerned. The most likely areas of wear are on the tangs of the drive gear or on the gear teeth. If wear is encountered on the teeth of either gear it is advisable to renew both together. Note also the condition of the large oil seal set in the backplate. If this seal shows any sign of damage or deterioration it must be renewed to prevent grease from the drive assembly working through to the brake linings.

4   Reassemble the speedometer drive assembly in the reverse order

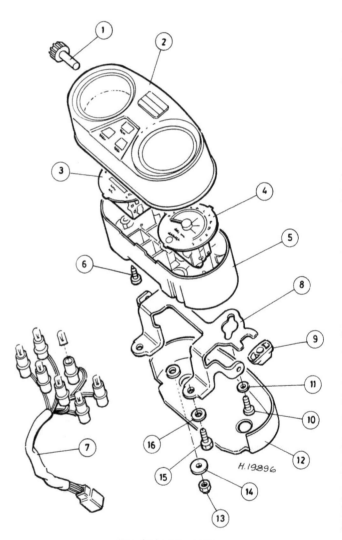

Fig. 4.6 Instruments

1 Reset knob	9 Mounting rubber – 2 off
2 Top cover	10 Screw – 4 off
3 Speedometer	11 Washer – 4 off
4 Tachometer	12 Panel base
5 Housing	13 Domed nut – 2 off
6 Screw – 2 off	14 Washer – 2 off
7 Bulbholder assembly	15 Bolt – 2 off
8 Mounting bracket	16 Washer – 2 off

17.2a Speedometer drive gear is located on inner face of brake backplate

17.2b Remove grub screw to free bush, driven gear and washers

17.5a Release circlip, thrust washer and retainer to remove tachometer drive gears

17.5b Driven gear can be withdrawn after grub screw has been removed

of dismantling. Lightly lubricate all components with a good quality high melting-point grease.

5   The tachometer drive assembly is located inside the crankcase right-hand cover and it will be necessary therefore to remove the cover as described in Section 9 of Chapter 1 in order to gain access to the drive components. A shouldered shaft set in the crankcase cover carries two interconnected drive pinions which are retained by a circlip and thrust washer. The first of these pinions, which is constructed of white nylon, meshes with the crankshaft primary drive pinion while the second, a worm drive pinion, meshes with the cable drive gear itself. The driven gear rotates in a sleeve set in the crankcase cover and is retained by a hexagon-headed grub screw. An O-ring is set in the sleeve to prevent oil leaks.

6   The mechanism is robust and well lubricated, and should require little attention. In the rare event of damage or wear occurring to any of the components, it will be necessary to use a pair of circlip pliers to remove the circlip which retains the two drive pinions, whereupon the thrust washer, retaining washer and the two pinions can be withdrawn from the shaft. The cable driven gear is removed by unscrewing the

retaining grub screw together with its washer, and withdrawing the drive gear and sleeve together. Any worn or damaged components should be renewed and the assembly rebuilt in the reverse of the dismantling order. Always fit a new sealing O-ring to the cable driven gear sleeve to prevent oil leaks.

18   Dualseat: removal and refitting

1   The dualseat is retained by a hooked plate at the front and a striker and catch assembly at the rear. A key-operated lock releases the catch via a short cable, enabling the seat to be removed.

2   Refit the seat by inserting its hook into the seat holding bracket on the frame and pushing down on the rear of the seat so that the strikers snap into their respective catches.

3   If the seat cover is damaged during an accident or split through old age, it should be renewed as soon as possible to prevent rusting of the

seat pan due to the entry of moisture and to preserve the machine's appearance. Seat covers can be obtained from manufacturers who advertise in the national motorcycle press, or a local motorcycle dealer may be able to offer a seat re-covering service.

4   The lock should be lubricated at regular intervals to offset the effects of corrosion. A water-dispersant spray applied via an aerosol can is the most convenient method of achieving this. Should the lock malfunction, it will be necessary to renew it as no repairs are possible. Be careful to carry a matching key when the machine is in use.

19   Frame covers and sidepanels: removal and refitting

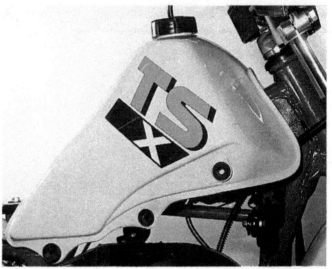

1   Each sidepanel is retained by a single screw plus a push fitting on the rear edge of the fuel tank and a hook which passes over the frame top rail. Remove the single crosshead screw and gently prise the front mounting out of its grommet in the fuel tank. Unhook the rear mounting and withdraw the sidepanel, taking care not to break the plastic.

2   The front frame covers incorporate a similar fitting arrangement. Separate the lugs from their respective grommets on the fuel tank and lift the covers off their lower mounts.

3   The plastic imitation radiator guard is retained by two crosshead screws to the frame.

19.1 Sidepanels and frame front covers push into grommets set in fuel tank

Chapter 5 Wheels, brakes and tyres

Contents

Specifications

Wheels

Size:	
Front	1.40 x 21
Rear...	1.60 x 18
Rim runout limit – axial and radial	2 mm (0.08 in)
Spindle runout limit...	0.25 mm (0.01 in)

Brakes

Drum inside diameter service limit:	
Front	120.7 mm (4.7520 in)
Rear...	100.7 mm (3.9646 in)
Brake shoe friction material service limit.....................	1.5 mm (0.0590 in)

Final drive

Chain size...	420 ($\frac{1}{2} \times \frac{1}{4}$)
Number of links...	124
20 pitch length service limit.......................................	259 mm (10.2 in)
Reduction ratio (no of teeth)	4.25 : 1 (51/12T)

Tyres

Size:	
Front	2.50 x 21 – 4PR
Rear...	3.00 x 18 – 4PR
Pressures (cold):	
Front	1.5 kg/cm² (21 psi)
Rear – solo...	1.75 kg/cm² (25 psi)
Rear – pillion..	2.00 kg/cm² (28 psi)
Manufacturer's recommended minimum tread depth	3.0 mm (0.04 in)

Torque wrench settings

Component	kgf m	lbf ft
Front wheel spindle nut..	2.7 – 4.3	20 – 31
Rear wheel spindle nut...	3.6 – 5.2	26 – 38
Brake torque arm nuts	1.0 – 1.5	7 – 11
Brake camshaft operating arm nut..............................	0.5 – 0.8	3.5 – 6
Rear sprocket retaining nut..	1.8 – 2.8	13 – 20

1 General description

All models are fitted with wheels of a traditional design, in which a full-width alloy hub is laced to a steel or alloy rim by means of steel spokes. All tyres are of the conventional tubed type.

Brakes are of the single leading shoe drum type operating inside a full width hub, the front being cable operated and the rear rod operated.

2 Front wheel: removal and refitting

1 Support the machine securely on a strong wooden box or similar support placed underneath the engine/gearbox unit so that the front wheel is raised clear of the ground. Slacken the brake cable retaining clamp situated at the top of the fork lower leg to permit the cable to be released, then slacken as far as possible the upper adjusting nut to give

2.2a Disconnect the speedometer drive cable

2.2b Remove the spindle nut ...

2.2c ... and pull the wheel spindle out from the left-hand side

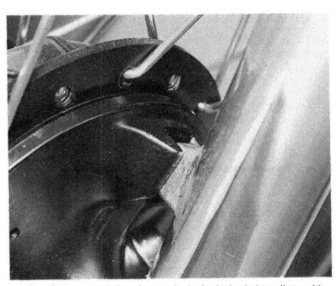

2.3 On refitting ensure that slot cast in the brake backplate aligns with lug on left-hand fork leg

the maximum cable free play. Slide the rubber sleeve down clear of the adjuster threads and unscrew the lower locking nut. Disengage the cable end nipple from the brake operating arm and slide the cable out of the adjuster stop cast in the brake backplate.

2   Disconnect the speedometer cable by slackening its lower knurled retaining ring with a suitable pair of pliers. Slacken and remove the wheel spindle retaining nut and remove the spindle by tapping it through the wheel and forks using a suitably sized long metal drift. Carefully pull out the spindle and lower the wheel to the ground.

3   On refitting the front wheel, ensure that the brake backplate is replaced so that the two tangs of the speedometer drive gear are correctly located in the slots provided in the hub, and that the spacer is pushed firmly into the right-hand side of the hub. Ensure that the wheel spindle is clean and completely free from corrosion, then smear high melting-point grease along its length to prevent the build-up of corrosion in the future. Offer the wheel assembly up to the forks ensuring that the raised lug on the fork left-hand lower leg engages correctly between the two raised ribs cast in the brake backplate. Push the spindle through from left to right, replace the spindle retaining nut, then tighten the nut by hand only at this stage.

4   Connect the speedometer driven cable, rotating the front wheel to ensure that the cable end meshes correctly with the driven pinion. Tighten securely the knurled retaining ring. Connect the front brake cable by reversing the cable removal procedure, then spin the front wheel and apply the front brake lever sharply. This will centralise the

shoes on the drum to achieve maximum braking efficiency. Maintain pressure on the front brake lever, clamping the shoes against the drum, and tighten the wheel spindle retaining nut to the specified torque setting. Adjust the front brake as described in Routine maintenance and secure the cable using the clamps provided.

5   Before taking the machine out on the road, check that the front wheel rotates easily and that the speedometer and the front brake are working correctly.

3 Rear wheel: removal and refitting

1   Raise the wheel clear of the ground by supporting the machine securely on a strong wooden box or similar support placed underneath the engine/gearbox unit.

2   Remove the split pin which secures the brake torque arm retaining nut. Remove the nut and spring washer and pull the torque arm off its mounting bolt set in the brake backplate. Detach the brake operating rod by unscrewing the adjusting nut from the threaded end of the rod and then depressing the brake pedal so that the rod is pulled clear of the brake operating arm; the nut can then be refitting to the rod for safekeeping.

3   Remove the nut, washer and chain adjuster indicator from the spindle. Slacken the drive chain adjusting nuts allowing the wheel to

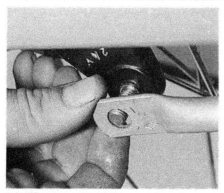

3.2a Disconnect the torque arm from the brake backplate ...

3.2b ... unscrew and remove the brake adjusting nut ...

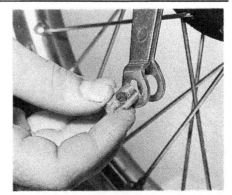

3.2c ... and disconnect the rod from the brake arm. Remove the trunnion for safekeeping

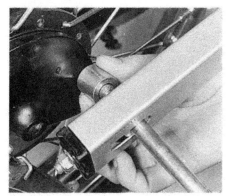

3.8a Insert the spindle through the swinging arm ends and wheel, not forgetting the spacers

3.8b Fit the chain adjuster indicator, washer and nut

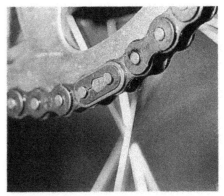

3.9a Ensure final drive chain connecting link closed end faces normal direction of chain travel

move forward easing the tension on the chain.

4   Disconnect the final drive chain via its connecting link and hang it over the swinging arm to keep it clear of any dirt or debris on the ground.

5   Using a long metal drift, tap out the spindle and catch the right-hand spacer as it drops clear. The wheel can then be withdrawn from the swinging arm.

6   Remove the rear brake backplate from the right-hand side of the hub and the spacer and cush drive assembly from the left- hand side of the hub. It is advisable to refit all washers, spacers and other wheel components back onto the spindle in the correct sequence as this will prevent their loss and serve as a guide to reassembly.

7   To refit the wheel, insert the brake backplate and cush drive assembly into the hub ensuring that the vanes on the sprocket carrier engage correctly with the rubber blocks set in the hub. Ensure that the left-hand chain adjuster indicator is correctly fitted at the head of the spindle. Check that the spindle, spacers and washers are free from dirt and corrosion and smear a small amount of high melting-point grease on the length of the spindle to prevent corrosion and ease any future removal and refitting of the wheel.

8   Push the spindle through the left-hand swinging arm fork and chain adjuster so that it protrudes only slightly. Offer up the wheel and push the spindle through the hub until it reaches the brake backplate. Fit the spacer and push the spindle fully into position, through the right-hand swinging arm fork and chain adjuster. Refit the right-hand chain adjuster indicator, washer and nut to the spindle, tightening the nut by hand at this stage.

9   Thread the final drive chain onto the rear sprocket and fit the connecting link. The closed end of the clip *must* face the direction of normal chain travel. Connect the brake torque arm to the brake back-plate, refit the spring washer and tighten the retaining nut to the specified torque setting. Secure the nut with a new split pin.

10   Refit the brake operating rod through the trunnion in the brake operating arm and refit the adjuster nut.

11   Check and if necessary adjust the chain tension, rear brake and rear stop lamp switch height as described in the relevant section of Routine maintenance. Tighten the rear spindle nut to the specified torque setting.

3.9b Use a new split pin to secure torque arm nut

12   Ensure that the rear brake and stop lamp are working correctly before using the machine on the road.

4 Wheels: examination and renovation

1   The wheel can be checked whilst in place on the machine after it has been raised clear of the ground. Make the machine as stable as possible, using blocks placed beneath the engine/gearbox unit as a support. Spin the wheel and ensure that there is no brake drag; if necessary, slacken the brake adjuster until the wheel turns freely. In the case of the rear

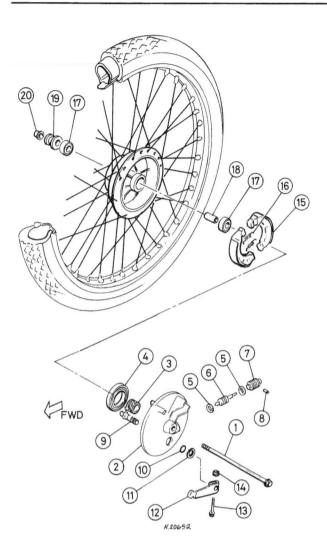

Fig. 5.1 Front wheel and brake

1 Wheel spindle	11 Washer
2 Brake backplate	12 Brake operating arm
3 Speedometer drive gear	13 Bolt
4 Oil seal	14 Nut
5 Washer – 2 off	15 Brake shoe – 2 off
6 Speedometer driven gear	16 Return spring – 2 off
7 Bush	17 Bearing – 2 off
8 Grub screw	18 Centre spacer
9 Brake camshaft	19 Right-hand spacer
10 O-ring	20 Nut

wheel, it is advisable, though not essential, to remove the final drive chain.

2   Slowly rotate the wheel and examine the rim for signs of serious corrosion or impact damage. Slight deformities, as might be caused by running the wheel along a curb, can often be corrected by adjusting spoke tension. More serious damage may require a new rim to be fitted, and this is best left to an expert. Whilst this is not an impossible undertaking at home, there is an art to wheel building, and a professional wheel builder will have the facilities and parts required to carry out the work quickly and economically. In the case of steel rims, if badly rusted they should be renewed in the interests of safety as well as appearance.

3   Assuming the wheel to be undamaged it will be necessary to check it for runout. This is best done by arranging a temporary wire pointer so that it runs close to the rim. The wheel can now be turned and any distortion noted. Check for lateral distortion and for radial distortion, noting that the latter is less likely to be encountered if the wheel was set

up correctly from new and has not been subject to impact damage.

4   The rim should be no more than 2.0 mm (0.08 in) out of true in either plane. If a significant amount of distortion is encountered check that the spokes are of approximately equal tension. Adjustment is effected by turning the square-headed spoke nipples with the appropriate spoke key. This tool is obtainable from most good motorcycle shops or tool retailers.

5   With the spokes evenly tensioned, any remaining distortion can be pulled out by tightening the spokes on one side of the hub and slackening the corresponding spokes from the opposite hub flange. This will allow the rim to be pulled across whilst maintaining spoke tension.

6   If more than slight adjustment is required it should be noted that the tyre and inner tube should be removed first to give access to the spoke ends. Those which protrude through the nipple after adjustment should be filed flat to avoid the risk of puncturing the tube. It is essential that the rim band is in good condition as an added precaution against chafing. In an emergency, use a strip of duct tape as an alternative: unprotected tubes will soon chafe on the nipples.

7   Should a spoke break, a replacement item can be fitted and retensioned in the normal way. Wheel removal is necessary for this operation, although complete removal of the tyre can be avoided if care is taken. A broken spoke should be attended to promptly because the load normally taken by that spoke is transferred to adjacent spokes which may fail in turn.

8   Remember to check wheel condition regularly. Normal maintenance is confined to keeping the spokes correctly tensioned and will avoid the costly and complicated wheel rebuilds that will inevitably result from neglect. When cleaning the machine do not neglect the wheels. If the rims are kept clean and well polished many of the corrosion related maladies will be prevented.

9   On completion of wheel examination, remove the machine support and if necessary, adjust the brake as described in Routine maintenance. In the case of the rear wheel reconnect the final drive chain.

5   Front wheel bearings: removal, examination and refitting

1   Remove the front wheel as described in Section 2. Withdraw the brake assembly and right-hand spacer from the hub.

2   Position the wheel on a work surface with its hub well supported by wooden blocks so that enough clearance is left beneath the wheel to drive the bearing out. Ensure the blocks are placed as close to the bearing as possible, to lessen the risk of distortion occurring to the hub casting whilst the bearings are being removed or fitted.

3   Place the end of a long-handled drift against the upper face of the lower bearing and tap the bearing downwards out of the wheel hub. The spacer located between the two bearings may be moved sideways slightly in order to allow the drift to be positioned against the face of the bearing. Move the drift around the face of the bearing whilst drifting it out of position, so that the bearing leaves the hub squarely.

4   With the one bearing removed, the wheel may be lifted and the spacer withdrawn from the hub. Invert the wheel and remove the second bearing, using a similar procedure to that used for the first.

5   Remove all the old grease from the hub and bearings, giving the latter a final wash in solvent. Check the bearings for signs of play or roughness when they are turned. If there is any doubt about the condition of a bearing, it should be renewed.

6   If the original bearings are to be refitted, then they should be repacked with high melting-point grease before being fitted into the hub. New bearings must also be prepared in the same way. Ensure that the bearing recesses in the hub are clean and both bearings and recess mating surfaces lightly greased to aid fitting. Check the condition of the hub recesses for evidence of abnormal wear which may have been caused by the outer race of a bearing spinning. If evidence of this happening is found, and the bearing is a loose fit in the hub, then it is best to seek advice from an authorized Suzuki dealer or a competent motorcycle engineer. Alternatively a proprietary product such as Loctite Bearing Fit may be used to retain the bearing outer race; this will mean, however, that the bearing housing must be carefully cleaned and degreased before the locking compound can be used.

7   With the wheel hub and bearing thus prepared, proceed to fit the bearings and central spacer as follows. With the hub again well supported by the wooden blocks, drift the first of the two bearings into position. To do this, use a soft-faced hammer in conjunction with a

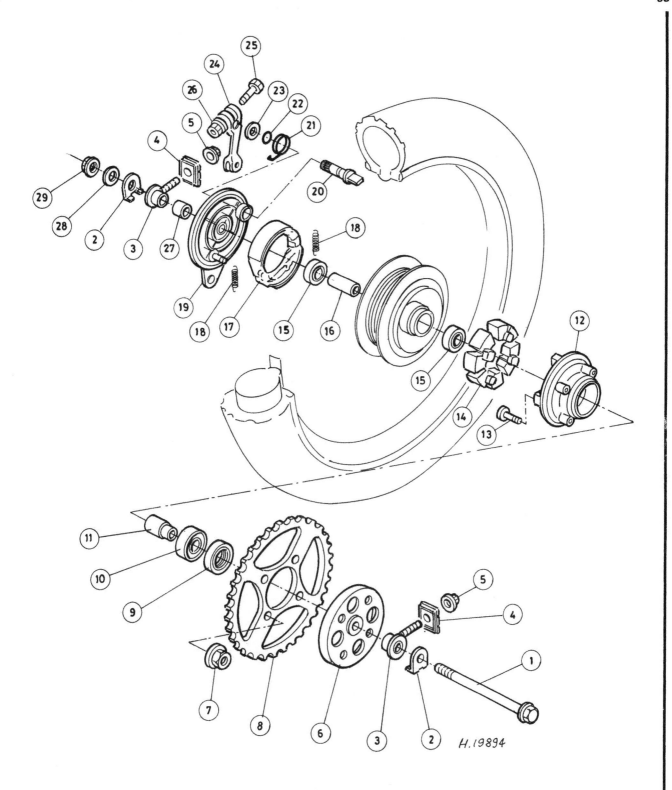

Fig. 5.2 Rear wheel and brake

H.19894

1	Wheel spindle	9	Oil seal	17	Brake shoe – 2 off	25	Bolt
2	Chain adjuster indicator – 2 off	10	Bearing	18	Return spring – 2 off	26	Nut
3	Chain adjuster – 2 off	11	Shouldered spacer	19	Brake backplate	27	Right-hand spacer
4	End plate – 2 off	12	Cush drive hub/sprocket carrier	20	Brake camshaft	28	Washer
5	Adjuster nut – 2 off	13	Bolt – 4 off	21	Return spring	29	Nut
6	Left-hand spacer	14	Cush drive rubbers – 4 off	22	O-ring		
7	Nut – 4 off	15	Bearing – 2 off	23	Washer		
8	Sprocket	16	Centre spacer	24	Brake operating arm		

5.3 Tap the bearing downwards out of its housing using a long-handled drift

socket or length of metal tube which has an overall diameter which is slightly less than that of the outer race of the bearing, but which does not bear at any point on the bearing sealed surface or inner race. Tap the bearing into place against the locating shoulder machined in the hub, remembering that the sealed surface of the bearing must always face outwards. With the first bearing in place, invert the wheel, insert the central spacer and pack the hub centre no more than $\frac{2}{3}$ full with high melting-point grease. Proceed to fit the second bearing, using the same procedure as given for the first. Take great care to ensure that each of the bearings enters its housing correctly, that is, square to the housing, otherwise the housing surface may be broached.

8   Refit the wheel to the machine as described in Section 2.

6   Rear wheel bearings: removal, examination and refitting

1   Before the rear wheel bearings can be examined, the wheel must first be removed from the machine. Refer to Section 3 of this chapter. Access to the bearings can be gained by removing the brake backplate and sprocket/cush drive assembly.
2   The rear hub assembly is essentially the same as that fitted to the front wheel and reference can be made to Section 5 for bearing removal, examination and refitting techniques.
3   The cush drive assembly carries an additional bearing for which removal, examination and refitting details can be found in the following section.

7   Rear cush drive: removal, examination and refitting

1   If any work is to be carried out to the rear cush drive assembly, the rear wheel must first be removed from the machine. Refer to Section 3 of this Chapter.
2   The rear sprocket is mounted on a separate carrier which rotates on the wheel spindle by means of a single journal ball bearing. Four large vanes on the rear face of the carrier engage with the cush drive rubber blocks set in the hub. This system allows the sprocket some movement within controlled limits and absorbs the shock loads produced from the final drive.
3   The sprocket is separated from the carrier by removing the four locknuts which are exposed once the left-hand spacer is removed. The sprocket need only be renewed if its teeth are hooked, chipped or badly worn. Comparison with a new component will give a good idea of the degree of wear but remember that it is bad practice to renew just one sprocket or to fit a new chain to part worn sprockets. This is because the running together of old and new components produces a far higher rate of wear than normal. The gearbox and rear wheel sprocket and the chain should always be renewed as a complete set.
4   The carrier bearing is removed by pulling out the shouldered spacer and drifting out the bearing using a hammer and socket or tubular drift

7.3 Remove spacer and four nuts to separate sprocket from cush drive hub

7.6 Cush drive rubber blocks should be renewed if they show signs of deterioration

7.8a Pack the carrier bearing with grease and drive into its housing

7.8b Use a similar method to fit the oil seal

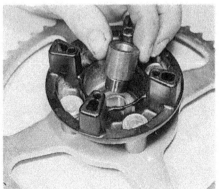

7.9a Insert the shouldered spacer ...

7.9b ... and insert the cush drive hub into the wheel

as described in Section 5 of this chapter. The oil seal will be displaced with the bearing or it can be levered out using a flat-bladed screwdriver.

5 Examine the carrier bearing as described in Section 5 of this chapter and renew it if necessary. The oil seal should be closely inspected for any indication of damage, hardening or perishing and renewed if necessary. It is advisable to renew this seal as a matter of course, particularly if the bearing is found to be defective.

6 The only other components which are likely to require renewal are the rubber blocks. Check carefully that the rubber is not split or breaking up and check that, when reassembled, excessive movement is not apparent. If any fault is found the rubber blocks must be renewed as a set. Note that roughness or harshness in the transmission, particularly when under acceleration or deceleration, indicates worn cush drive rubbers.

7 Reassembly is a straightforward reversal of the dismantling procedure, noting the following points. Lubricate the cush drive rubber blocks with a solution of soapy water or with a water dispersant fluid such as WD40; this will make the surface of the rubber slippery, thus ensuring that the blocks slide easily into place.

8 Pack the carrier bearing with high melting-point grease and drift into place. Fit the oil seal against the bearing using a similar method.

9 On fitting the rear sprocket ensure that new locknuts are fitted and tighten the nuts to the specified torque setting. Insert the cush drive assembly into the wheel hub not forgetting to insert the spacers.

10 Insert the brake backplate in the hub right-hand side and refit the rear wheel to the machine as described in Section 3 of this chapter.

8 Front brake: examination and renovation

1 The brake assembly, complete with the brake backplate, can be withdrawn from its wheel hub after removal of the wheel from the machine. With the wheel laid on a work surface, brake backplate uppermost, the brake backplate may be lifted away from the hub. It will come away quite easily, with the brake shoe assembly attached to its back.

2 Examine the condition of the brake linings. If they are worn beyond

the service limit then the brake shoes should be renewed. The linings are bonded on and cannot be supplied separately.

3 If oil or grease from the wheel bearings has badly contaminated the linings, the brake shoes should be renewed. There is no satisfactory way of degreasing the lining material. Any surface dirt on the linings can be removed with a stiff-bristled brush. High spots on the linings should be carefully eased down with emery cloth.

4 Examine the drum surface for signs of scoring, wear beyond the service limit or oil contamination. All of these conditions will impair braking efficiency. Remove all traces of dust, preferably using a brass wire brush, taking care not to inhale any of it, as it is of an asbestos nature, and consequently harmful. Remove oil or grease deposits, using solvent.

5 If deep scoring is evident, due to the linings having worn through to the shoe at some time, the drum must be skimmed on a lathe, or renewed. Whilst there are firms who will undertake to skim a drum whilst it is fitted to the wheel, it should be borne in mind that excessive skimming will change the radius of the drum in relation to the brake shoes, therefore reducing the friction area until extensive bedding in has taken place. Also full adjustment of the shoes may not be possible. If in doubt about this point, the advice of one of the specialist engineering firms who undertake this work should be sought.

6 Note that it is false economy to try to cut corners with brake components; the whole safety of both machine and rider being dependent on their good condition.

7 Removal of the brake shoes is accomplished by folding the shoes together so that they form a 'V'. With the spring tension relaxed, both shoes and springs may be removed from the brake backplate as an assembly. Detach the springs from the shoes and carefully inspect them for any signs of fatigue or failure. If in doubt, compare them with a new set of springs.

8 Before fitting the brake shoes, check that the brake operating camshaft is working smoothly and is not binding in its pivot. The camshaft can be removed by withdrawing the retaining bolt on the operating arm and pulling the arm off the shaft. Before removing the arm, it is advisable to mark its position in relation to the shaft, so that it can be relocated correctly.

9 Remove any deposits of hardened grease or corrosion from the bearing surface of the brake cam by rubbing it lightly with a strip of fine

8.2 Measure the friction material thickness

8.4 Examine the drum surface for signs of scoring and check its internal diameter

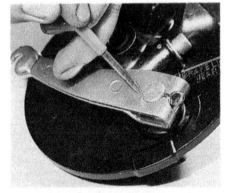
8.8a Mark the brake arm position in relation to the camshaft before removal

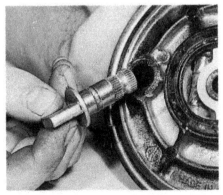
8.8b Withdraw the camshaft from the brake backplate

8.10a When refitting the camshaft insert the O-ring ...

8.10b ... and washer to its external end

emery paper or by applying solvent with a piece of rag. Apply a smear of high melting-point grease to the length of the shaft and the face of the operating cam prior to reassembly. Clean and lightly grease the pivot stub which is set in the backplate.

10   Check the condition of the O-ring which prevents the escape of grease from the end of the camshaft. If it is in any way damaged or perished, then it must be renewed before the shaft is relocated in the backplate. Relocate the camshaft and align and fit the operating arm with the O-ring and plain washer. The bolt and nut retaining the arm in position on the shaft should be tightened to the specified torque setting.

11   Before refitting existing shoes, roughen the lining surface sufficiently to break the glaze which will have formed in use. Glasspaper or emery cloth is ideal for this purpose but take care not to inhale any of the asbestos dust that may come from the lining surface.

12   Fitting the brake shoes and springs to the brake backplate is a reversal of the removal procedure. Some patience will be needed to align the assembly with the pivot and operating cam whilst still retaining the springs in position; once they are correctly aligned though, they can be pushed back into position by pressing downwards in order to snap them into position. Do not use excessive force, or there is risk of distorting the brake shoes permanently.

13   Adjustment of the front brake is fully described in the relevant section of Routine maintenance. Take great care that the brake functions efficiently before using the machine on the road.

9 Rear brake: examination and renovation

1   If any work is to be carried out on the rear brake, the rear wheel must first be removed from the machine as described in Section 3. The rear brake is essentially similar to the unit at the front and reference can therefore be made to Section 8 of this Chapter for details of examination and renovation techniques. Note, however, the following differences which should be taken into consideration when working on the rear brake assembly.

2   An external return spring is fitted to the rear brake operating arm to assist the release of the brake. The spring can be removed complete with the operating arm but care should be taken on reassembly to fit the spring in the correct position.

3   If the shoes or their return springs require renewal, note that during the XKE model production run these components were modified. The return spring length is 31.2 mm on early models and 33 mm on all later models; ensure that the correct parts are obtained if renewal is required.

4   After inspection of the rear brake, refit the wheel to the machine (Section 3) and adjust the operating mechanism (Routine maintenance). Ensure that the brake is working correctly before taking the machine out on the road.

10 Final drive chain: examination, cleaning and lubrication

1   The final drive chain is fully exposed apart from the protection given by a short chainguard along the upper run, and if not properly maintained will have a short life. A worn chain will cause rapid wear of the sprockets and they too will need renewal.

2   A simple check can be carried out to assess the amount of wear that has taken place in the chain. With the chain fully lubricated and adjusted as described in the Routine maintenance Section, attempt to pull the chain backwards off the rear sprocket. If the chain can be pulled clear of the sprocket teeth it must be considered worn out and should be renewed, in conjunction with the sprockets. A more accurate measurement of chain wear involves removing the chain from the machine.

3   Disconnect the chain at its split connecting link and pull the entire length clear of the sprockets. Note that refitting the chain is greatly simplified if a worn-out length is temporarily connected to it. As the original chain is pulled off the sprockets, the worn-out chain will follow it and remain in place while the task of cleaning and examination is carried out. On reassembly, the process is repeated, pulling the worn-out chain over the sprockets so that the new chain, or the freshly cleaned and lubricated chain, is pulled easily into place.

4   To clean the chain, immerse it in a bath containing a high flash-point solvent and use a stiff-bristled brush to scrub away all traces of road dirt

and old lubricant. Swill the chain around to ensure that the solvent penetrates fully into the bushes and rollers and can remove any lubricant which may still be present. When the chain is completely clean, remove it from the bath and hang it up to dry.

5   To assess accurately the amount of wear present in the chain, it must be cleaned and dried as described above, then laid out on a flat surface. Compress the chain fully and measure its length from end to end. Anchor one end of the chain and pull on the other end, drawing the chain out to its fullest extent. Measure the stretched length. If the stretched measurement exceeds the compressed measurement by more than $\frac{1}{4}$ in per foot, the chain must be considered worn out and be renewed. Suzuki's own recommendation is that a 20 pitch length, when cleaned and dried, must not exceed 259 mm (10.2 in). This is accomplished by marking any pin along the length of the chain, counting off 21 pins, and by measuring the distance between the two.

6   Periodic chain lubrication should be made as described in Routine maintenance at the specified intervals. A thorough lubrication of the chain is, however, best carried out by immersing the chain in a molten lubricant such as Chainguard or Linklyfe. Lubrication carried out in this manner must be preceded by removing the chain from the machine, cleaning it, and drying it as described above. Follow the manufacturer's instructions carefully when using Chainguard or Linklyfe, and take great care to swill the chain gently in the molten lubricant to ensure that all bearing surfaces are fully greased.

7   Refitting a new, or freshly-lubricated, chain is a potentially messy affair which is greatly simplified by the substitution of a worn-out length of chain during removal. The new chain can then be connected to the worn-out length and pulled easily around the sprockets. Refit the connecting link, ensuring that the spring clip is fitted with its closed end facing the normal direction of travel of the chain.

8   When ordering a new chain, always quote the size, number of chain links and the full model name.

11 Tyres: removal, repair and refitting

1   To remove the tyre from either wheel, detach the wheel from the machine. Deflate the tyre by removing the valve core, and when the tyre is fully deflated, push the bead away from the wheel rim on both sides so that the bead enters the centre well of the rim. Remove the locking ring and push the tyre valve into the tyre itself.

2   Insert a tyre lever close to the valve and lever the edge of the tyre over the outside of the rim. Very little force should be necessary; if resistance is encountered it is probably due to the fact that the tyre beads have not entered the well of the rim, all the way round. If aluminium rims are fitted, damage to the soft alloy by tyre levers can be prevented by the use of plastic rim protectors.

3   Once the tyre has been edged over the wheel rim, it is easy to work round the wheel rim, so that the tyre is completely free from one side. At this stage the inner tube can be removed.

4   Now working from the other side of the wheel, ease the other edge of the tyre over the outside of the wheel rim that is furthest away. Continue to work around the rim until the tyre is completely free from the rim.

5   If a puncture has necessitated the removal of the tyre, reinflate the inner tube and immerse it in a bowl of water to trace the source of the leak. Mark the position of the leak, and deflate the tube. Dry the tube, and clean the area around the puncture with a high flash-point solvent. When the surface has dried, apply rubber solution and allow this to dry before removing the backing from the patch, and applying the patch to the surface.

6   It is best to use a patch of self-vulcanizing type, which will form a permanent repair. Note that it may be necessary to remove a protective covering from the top surface of the patch after it has sealed into position. Inner tubes made from a special synthetic rubber may require a special type of patch and adhesive, if a satisfactory bond is to be achieved.

7   Before refitting the tyre, check the inside to make sure that the article that caused the puncture is not still trapped inside the tyre. Check the outside of the tyre, particularly the tread area to make sure nothing is trapped that may cause a further puncture.

8   If the inner tube has been patched on a number of past occasions, or if there is a tear or large hole, it is preferable to renew it. Sudden deflation may cause an accident, particularly if it occurs with the rear wheel.

Tyre changing sequence - tubed tyres

 A Deflate tyre. After pushing tyre beads away from rim flanges push tyre bead into well of rim at point opposite valve. Insert tyre lever adjacent to valve and work bead over edge of rim.

Use two levers to work bead over edge of rim. Note use of rim protectors **B**

Remove inner tube from tyre **C**

When first bead is clear, remove tyre as shown **D**

When fitting, partially inflate inner tube and insert in tyre **E**

Work first bead over rim and feed valve through hole in rim. Partially screw on retaining nut to hold valve in place. **F**

 G Check that inner tube is positioned correctly and work second bead over rim using tyre levers. Start at a point opposite valve.

Work final area of bead over rim whilst pushing valve inwards to ensure that inner tube is not trapped **H**

9   To refit the tyre, inflate the inner tube for it just to assume a circular shape but only to that amount, and then push the tube into the tyre so that it is enclosed completely. Lay the tyre on the wheel at an angle, and insert the valve through the rim tape and the hole in the wheel rim. Attach the locking ring on the first few threads, sufficient to hold the valve captive in its correct location.

10   Starting at the point furthest from the valve, push the tyre bead over the edge of the wheel rim until it is located in the central well. Continue to work around the tyre in this fashion until the whole of one side of the tyre is on the rim. It may be necessary to use a tyre lever during the final stages.

11   Make sure there is no pull on the tyre valve and again commencing with the area furthest from the valve, ease the other bead of the tyre over the edge of the rim. Finish with the area close to the valve, pushing the valve up into the tyre until the locking ring touches the rim. This will ensure that the inner tube is not trapped when the last section of bead is edged over the rim with a tyre lever.

12   Check that the inner tube is not trapped at any point. Reinflate the inner tube, and check that the tyre is seating correctly around the wheel rim. There should be a thin rib moulded around the wall of the tyre on both sides, which should be an equal distance from the wheel rim at all points. If the tyre is unevenly located on the rim, try bouncing the wheel when the tyre is at the recommended pressure. It is probable that one of the beads has not pulled clear of the centre well.

13   Tyre replacement is aided by dusting the side walls, particularly in the vicinity of the beads, with a liberal coating of french chalk or a proprietary tyre fitting lubricant.

14   Never replace the inner tube and tyre without the rim tape in position. If this precaution is overlooked there is a good chance of the ends of the spoke nipples chafing the inner tube and causing a crop of punctures.

15   Never fit a tyre that has a damaged tread or sidewalls. Apart from legal aspects, there is a very great risk of a blowout, which can have very serious consequences on a two wheeled vehicle.

16   Always run the tyres at the correct pressures and never under or over inflate. The correct pressures are given in the Specifications section of this chapter. If non standard tyres are fitted, check with the tyre manufacturer or supplier for recommended pressures.

12   Valve cores and caps: general

1   Valve cores seldom give trouble, but do not last indefinitely. Dirt under the seating will cause a puzzling 'slow- puncture'. Check that they are not leaking by applying spittle to the end of the valve and watching for air bubbles.

2   A valve cap is a safety device, and should always be fitted. Apart from keeping dirt out of the valve, it provides a second seal in case of valve failure, and may prevent an accident resulting from sudden deflation.

Chapter 6 Electrical system

Contents

Specifications

Battery

Make..	Yuasa
Type	YB4L-B
Capacity	4Ah
Earth..	Negative
Electrolyte specific gravity	1.280 @ 20°C (68°F)

Fuse

Rating..	15 amp

Flywheel generator

Charging coil resistance (yellow to yellow)............................	0.5 – 0.8 ohm
Charging system regulated output..	14 – 15V @ 5000 rpm
Generator no-load voltage..	40V (ac) minimum @ 5000 rpm

Bulbs

Headlamp..	12V, 35/35W
Parking lamp	12V, 4W
Stop/tail lamp...	12V, 21/5W
Turn signal lamp...	12V, 21W
Oil level warning lamp	12V, 3.4W
Neutral indicator lamp...	12V, 3.4W
High beam warning lamp	12V, 1.7W
Turn signal warning lamp	12V, 3.4W
Instrument illuminating lamps.......................................	12V, 1.7W

1 General description

The 12 volt electrical system is powered by a crankshaft-mounted flywheel generator, located behind the left-hand crankcase cover. The generator contains four coils, one supplying power for the ignition circuit and the other three providing charge for the battery. The ignition circuit and charging coils are in no way connected and for the purpose of testing can be considered as separate components. Refer to Chapter 3 for details of the ignition source coil.

Alternating circuit (ac) from the generator, once converted to direct current (dc) by the rectifier unit, charges the battery. This unit also contains a voltage regulator which controls the level of charge, thus ensuring that the battery does not become overcharged.

2 Electrical system: testing – general

1 In the event of an electrical system fault, always check the physical condition of the wiring and connectors before attempting any of the test procedures described here and in subsequent Sections. Look for chafed, trapped or broken electrical leads and repair or renew these as necessary. Leads which have broken internally are not easily spotted, but may be checked using a multimeter or a simple battery and bulb circuit as a continuity tester. This arrangement is shown in the accompanying illustration. The various multi-pin connectors are generally trouble-free but may corrode if exposed to water. Clean them carefully, scraping off any surface deposits, and pack with silicone grease during assembly to avoid recurrent problems. The same technique can be applied to the

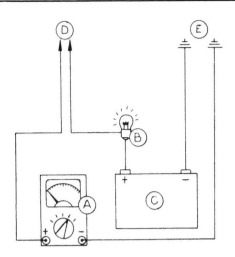

Fig. 6.1 Simple circuit testing equipment

A   Multimeter   D   Positive probe
B   Bulb   E   Negative probe
C   Battery

handlebar switches.

2   The wiring harness is colour-coded and will correspond with the wiring diagram at the end of this Manual. When socket connections are used, they are designed so that reconnection can be made only in the correct position.

3   Visual inspection will usually show whether there are any breaks or frayed outer coverings which will give rise to short circuits. Occasionally a wire may become trapped between two components, breaking the inner core but leaving the more resilient outer cover intact. This can give rise to mysterious intermittent or total circuit failure. Another source of trouble may be the snap connectors and sockets, where the connector has not been pushed fully home in the outer housing, or where corrosion has occurred.

4   Intermittent short circuits can often be traced to a chafed wire that passes through or is close to a metal component such as a frame member. Avoid tight bends in the lead or situations where a lead can become trapped between casings.

5   A sound, fully charged battery, is essential to the normal operation of the system. There is no point in attempting to locate a fault if the battery is partly discharged or worn out. Check battery condition and recharge or renew the battery before proceeding further.

6   Many of the test procedures described in this Chapter require voltages or resistances to be checked. This necessitates the use of some form of test equipment such as a simple and inexpensive multimeter of the type sold by electronics or motor accessory shops.

7   If you doubt your ability to check the electrical system, entrust the work to an authorised Suzuki dealer. In any event have your findings double-checked before consigning expensive components to the scrap bin.

3   Battery: examination and maintenance

1   Details of the regular checks needed to maintain the battery in good condition are given in Routine maintenance, together with instructions on removal and refitting and general battery care. Batteries can be dangerous if mishandled; read carefully the 'Safety first' section at the front of this Manual before starting work, and always wear overalls or old clothing in case of accidental acid spillage. If acid is ever allowed to splash into your eyes or onto your skin, flush it away with copious quantities of fresh water and seek medical advice immediately.

2   When new, the battery is filled with an electrolyte of dilute sulphuric acid having a specific gravity of 1.280 at 20°C (68°F). Subsequent evaporation, which occurs in normal use, can be compensated for by topping up with distilled or demineralised water only. Never use tap water as a substitute and do not add fresh electrolyte unless spillage has occurred.

3   The state of charge of a battery can be checked using an hydrometer.

4   The normal charge rate for a battery is 1/10 of its rated capacity, thus for a 4 ampere hour unit charging should take place at 0.4 amp. Exceeding this figure can cause the battery to overheat, buckling the plates and rendering it useless. Few owners will have access to an expensive current controlled charger, so if a normal domestic charger is used check that after a possible initial peak, the charge rate falls to a safe level. If the battery becomes hot during charging **stop**. Further charging will cause damage. Note that cell caps should be loosened and vents unobstructed during charging to avoid a build-up of pressure and risk of explosion.

5   After charging top up with distilled water as required, then check the specific gravity and battery voltage. Specific gravity should be above 1.270 and a sound, fully charged battery should produce 15 - 16 volts. If the recharged battery discharges rapidly if left disconnected it is likely that an internal short caused by physical damage or sulphation has occurred. A new battery will be required. A sound item will tend to lose its charge at about 1% per day.

4   Charging system: checking the output

1   Before commencing this test, it is essential that the battery is fully charged and in good condition. If this is not the case, the test results will be inconclusive. Remove the right-hand sidepanel to gain access to the battery.

2   Run the engine until it reaches its normal operating temperature. Note that this test involves running the engine at a fast speed so care must be taken to ensure that it does not overheat. Also the throttle should only be held at the specified engine speed long enough for the voltage reading to be taken.

3   With the engine stopped, connect a multimeter or dc voltmeter, set to the 0–20V dc range, as follows. Connect its positive (+) probe to the positive (+) battery terminal and the negative (-) probe to the negative (-) battery terminal. Start the engine and run it at 5000 rpm with the headlamp on and switched to the high beam position. The charging output is correct if the tester reads 14–15 volts dc. If the readings shown differ from this, check the flywheel generator charging coils and regulator/rectifier unit as described in the following sections.

5   Flywheel generator: charging coil test

1   If a fault is indicated in the charging system the charging coils should be checked as described below. It is not necessary to remove the flywheel generator to carry out this test.

2   Remove the seat and trace the main generator lead from the engine/gearbox unit up to its block connector. Separate the two connector halves and identify the two yellow wires on the generator feed side.

3   Connect a multimeter, set to the ohms x 1 range, between the two yellow wires. The reading taken should be as detailed in the Specifications section.

4   If this reading varies considerably, ie infinite or zero indicated resistance, it is safe to assume that an open or short circuit exists in the coils under test.

5   Before condemning an otherwise sound coil check the wires for breakage as described in Section 2 of this Chapter. If renewal is necessary the flywheel and stator coils can be removed as described in Chapter 1, Section 7. Note that it may be possible to repair any damage or even to have a coil rewound; always have your tests confirmed by an authorized Suzuki dealer or auto electrical expert and seek his advice as to whether repairs are possible or if a new unit is required.

6   If no resistance is shown in the coils, check that they are insulated from the stator so that no short circuiting against the engine is possible.

6   Regulator/rectifier unit: location and testing

1   The regulator/rectifier is a finned metal unit mounted at the top of

6.1 Regulator/rectifier unit location

7.1 Main fuse and spare are housed in plastic holder under right-hand sidepanel

the frame front downtubes. To test the unit, unplug its connector to expose the unit terminals.

2   With a multimeter set to its K ohm range, refer to the accompanying table and measure the resistance between the combinations of leads shown. Adhere strictly to the sequence given in the table and take great care not to allow the probes of the meter to short to earth or against each other. If, during the test sequence, the resistance readings shown on the meter scale differ from those given in the table, then the regulator/rectifier unit may be assumed to be defective and should be renewed.

7   Fuse: location and renewal

1   The electrical system is protected by a single 15 amp spade-type fuse which is located in a holder, inside the right-hand sidepanel, next to the battery. Unclip the fuse holder cover to gain access; the fuse is a simple push fit. If the spare fuse is used, replace it with one of the correct rating as soon as possible.

2   In the event of a fuse 'blowing', always try to find the cause of the fault before renewing it. If no obvious fault is noted, renew the fuse. If the new fuse operates normally, the system can be considered sound, but repeated failures indicate an intermittent fault, possibly due to a damaged wire or water contamination.

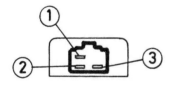

UNIT : KΩ

⊕ PROBE OF TESTER				
⊖ PROBE OF TESTER	1	2	3	GROUND
1		OFF	OFF	OFF
2	2 — 4		OFF	OFF
3	2 4	OFF		OFF
GROUND	6 —12	2—4	2 — 4	

H19895

Fig. 6.2 Regulator/rectifier unit test

8   Switches: general maintenance and repairs

1   While the switches should give little trouble, they can be tested using a multimeter set to the resistance function or a battery and bulb test circuit. Using the information given in the wiring diagram at the end of this Manual, check that full continuity exists in all switch positions and between the relevant pairs of wires. When checking a particular circuit, follow a logical sequence to eliminate the switch concerned.

2   As a simple precaution always disconnect the battery (negative lead first) before removing any of the switches, to prevent the possibility of a short circuit. Most troubles are caused by dirty contacts, but in the event of the breakage of some internal part, it will be necessary to renew the complete switch.

3   It should, however, be noted that if a switch is tested and found to be faulty, there is nothing to be lost by attempting a repair. It may be that worn contacts can be built up with solder, or that a broken wire terminal can be repaired, again using a soldering iron. The handlebar switches can all be dismantled to a greater or lesser extent. It is however, up to the owner to decide if he has the skill to carry out this sort of work.

4   While none of the switches require routine maintenance of any sort, some regular attention will prolong their life a great deal. In the author's experience, the regular and constant application of WD40 or a similar water-dispersant spray not only prevents problems occurring due to waterlogged switches and the resulting corrosion, but also makes the switches much easier and more positive to use. Alternatively, the switch may be packed with a silicone-based grease to achieve the same result.

5   The handlebar switches can be checked by tracing the main lead back to the appropriate block connector at the steering head. Use the wiring diagram at the back of this manual to identify the switch to be tested, and its colour-coded connecting wires. Set the multimeter to the resistance scale and place its probes on the connections of the switch in question. If continuity exists in all of the switches the entire unit can be deemed serviceable. However, if any of the tests show that one of the switches is faulty and no satisfactory repairs can be made, the unit must be renewed.

6   The neutral indicator switch takes the form of a black plastic cover which is sited over the left-hand end of the gearchange selector drum. Access to the switch may be gained by removing the crankcase left-hand cover.

7   To test the switch, set a multimeter to its resistance function and carry out a check for continuity between the switch terminal and earth with the gearchange lever set in the neutral position. If continuity is found, the switch is serviceable. If this is not the case, the switch should be removed by following the instructions given in Section 8 of Chapter 1. The switch components can be cleaned and examined, and any defective parts renewed. Switch reassembly is described in Section 35 of Chapter 1.

8   Two stop lamp switches are fitted to the machine described in this manual; the first being located in the front brake handlebar lever assembly and operated directly by the front brake lever, and the second being mounted on a bracket near the swinging arm pivot and operated by the rear brake pedal. Both switches can be checked for continuity by disconnecting the wires from the switch to be tested and connecting a multimeter set to the resistance x 1 ohm scale in their place. Apply the rear brake pedal or brake lever so that the switch is in the fully extended or on position. If no continuity is found the switch can be considered to be faulty and because it is a sealed unit, renewal is the only form of repair. The regular application of a water dispersant spray will prevent the onset of corrosion and prolong the life of the switches.

9   If a fault is suspected in the ignition switch, refer to Chapter 3, Section 4 for further details.

9   Oil level warning circuit: location and testing

1   If the instrument console mounted oil level warning bulb fails to illuminate when the ignition is switched to the C position, check first to see if the bulb has blown, following the instructions given in Section 14 of this Chapter.

2   The oil level warning sender unit is screwed into the top of the oil tank and access is made by removing the seat.

3   The sender unit is a simple float type, the contacts closing as the float drops down to its stop. Check continuity using a multimeter set on the resistance scale. If continuity is shown when the float is against the stop, with isolation being indicated when it is lifted, the sender unit can be considered sound.

4   If the sender unit is inoperative, try cleaning the contacts using fine abrasive paper. If this fails to rectify the problem, check that the leads are unbroken. As a last resort, renew the sender unit.

9.2 Oil level sender unit is housed in top of oil tank

10   Horn: location and testing

1   The horn is fixed to a metal bracket which is bolted to the frame front downtubes, on the right-hand side of the machine. No maintenance is required other than regular cleaning to remove road dirt and the occasional application of WD40, or a similar water dispersant lubricant, to minimise internal corrosion.

2   A locknut is provided on the outside of the horn so that the internal contacts may be adjusted to compensate for wear and to cure a weak or intermittent horn note. Slacken the locknut and rotate the screw slowly until the clearest and loudest note is obtained, then retighten the locknut. If no means of adjustment is provided on the horn fitted, it must be renewed.

3   If the horn fails to work, first check that power is reaching it by disconnecting the wires. Substitute a 12 volt bulb, switch on the ignition and press the horn button. If the bulb lights, the circuit is proved good and the horn is at fault; if the bulb does not light, there is a fault in the circuit which must be found and rectified.

4   To test the horn itself, connect a fully-charged 12 volt battery directly to the horn. If it does not sound, a gentle tap on the outside may free the internal contacts. If this fails, the horn must be renewed as repairs are not possible.

10.1 Horn is mounted via bracket, to frame front downtubes

11   Turn signals: testing and bulb renewal

1   In the event of a fault in the turn signals, it is essential that the problem is diagnosed methodically to avoid a good deal of confusion and wasted time. Start by attempting to operate both right and left turn signals in turn. If one side operates normally, the fault can only lie in the bulbs, wiring or switch contacts on the affected side; the relay can be taken as sound. The most likely cause by far is a blown bulb, and this should always be checked first. Usually, the remaining bulb on that side will flash dimly and rapidly, indicating the faulty item.

2   To gain access to the bulb, remove the two lens retaining screws and lift the lens away. The bulb is a bayonet fitting and is removed by

11.2 Turn signal bulbs are a bayonet fitting. Check lamp seal before refitting lens

11.4 Turn signal relay location

depressing it slightly and turning it anticlockwise. If the suspect bulb proves sound, check the contacts in the bulbholder, the wiring and the switch contacts, to locate the fault.

3   Where the fault is common to the whole turn signal circuit, attention should be directed to the turn signal relay and its wiring. Check the wiring first, and if this proves to be sound, check the relay by substituting a new one.
4   The relay is rubber mounted to the left-hand side of the frame next to the engine oil tank. Access can be gained simply by removing the seat.

12   Headlamp: bulb renewal and beam alignment

1   Remove the headlamp cowling by unscrewing its retaining screws. The headlamp bulb is retained to the reflector by a spring. Release the spring and withdraw the bulb and its holder from the reflector and lens. Press in the holder and turn it anticlockwise to release the bulb.
2   The parking lamp bulb assembly is a push fit into the headlamp reflector. The bulb should be pushed in and twisted anticlockwise to release it from its holder.
3   The headlamp beam is adjustable in both the vertical and horizontal planes. The upper adjuster controls horizontal alignment and the lower adjuster controls vertical alignment.
4   In the UK, regulations stipulate that the headlamp must be arranged so that the light will not dazzle a person standing at a distance greater than 25 feet from the lamp, whose eye level is not less than 3 feet 6 inches above that plane. It is easy to approximate this setting by placing the machine 25 feet away from a wall, on a level road, and setting the dip beam height so that it is concentrated at the same height as the distance

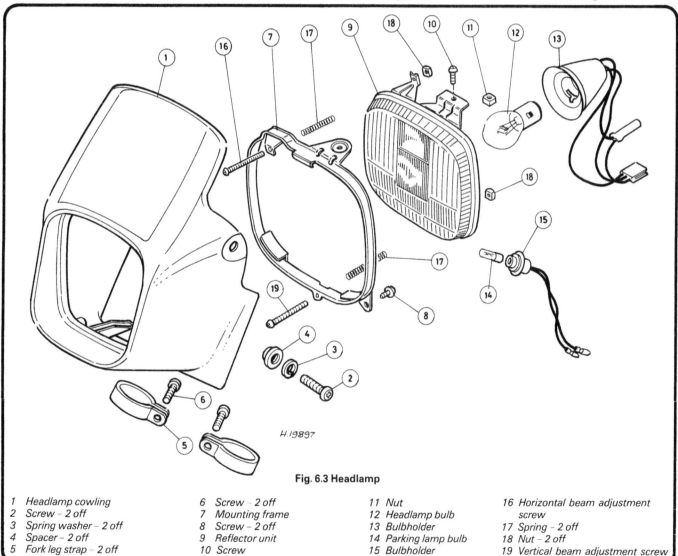

H.19897

Fig. 6.3 Headlamp

1 Headlamp cowling	6 Screw – 2 off	11 Nut
2 Screw – 2 off	7 Mounting frame	12 Headlamp bulb
3 Spring washer – 2 off	8 Screw – 2 off	13 Bulbholder
4 Spacer – 2 off	9 Reflector unit	14 Parking lamp bulb
5 Fork leg strap – 2 off	10 Screw	15 Bulbholder

16   Horizontal beam adjustment screw
17   Spring – 2 off
18   Nut – 2 off
19   Vertical beam adjustment screw

12.1a Headlamp cowling top mounting screws

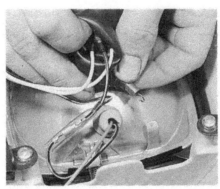

12.1b Release retaining spring and withdraw headlamp bulbholder

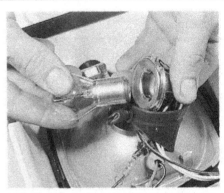

12.1c Push inwards and twist anticlockwise to separate headlamp bulb from holder

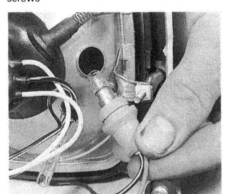

12.2 Parking lamp bulbholder is a push fit in reflector

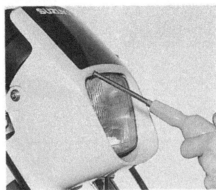

12.3a Headlamp beam is adjustable in horizontal ...

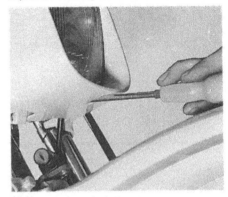

12.3b ... and vertical planes

of the centre of the headlamp from the ground. The rider must be seated normally during this operation and also the pillion passenger, if one is carried regularly.

2   Remove all traces of moisture and corrosion from the interior of the lamp and the bulbholder, and renew the sealing gasket if it is damaged. Refit the bulb, noting that the pins are offset so that the bulb can be fitted only the correct way. Do not overtighten the lens retaining screws or the lens may crack.

13   Stop/tail lamp: bulb renewal

1   Remove the two screws securing the lens and withdraw the lens, taking care not to tear the sealing gasket. Press the bulb in and twist it anticlockwise to release it.

14   Instrument panel: bulb renewal

1   Remove the headlamp cowling to expose the base of the instrument panel. Remove the two domed nuts from the panel base and slide

13.1a Stop/tail lamp lens is retained by two screws

13.1b Stop/tail lamp bulb is bayonet fitting and has offset pins

14.1 Remove instrument lower panel to gain access to bulbs

14.2 Instrument bulbholders are a push fit in housing

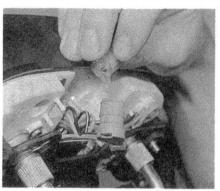

14.3 Take care not to damage bulbs on refitting

the cover down past the speedometer and tachometer drive cables.

2   Each bulbholder is a push fit into the instrument housing. The bulbs are of the capless type and should be pulled out of their respective holders.

3   Refitting is a reversal of the above, taking care not to damage the fine wire tails of the bulbs when refitting into their holders.

4   Details of instrument removal will be found in Chapter 4.

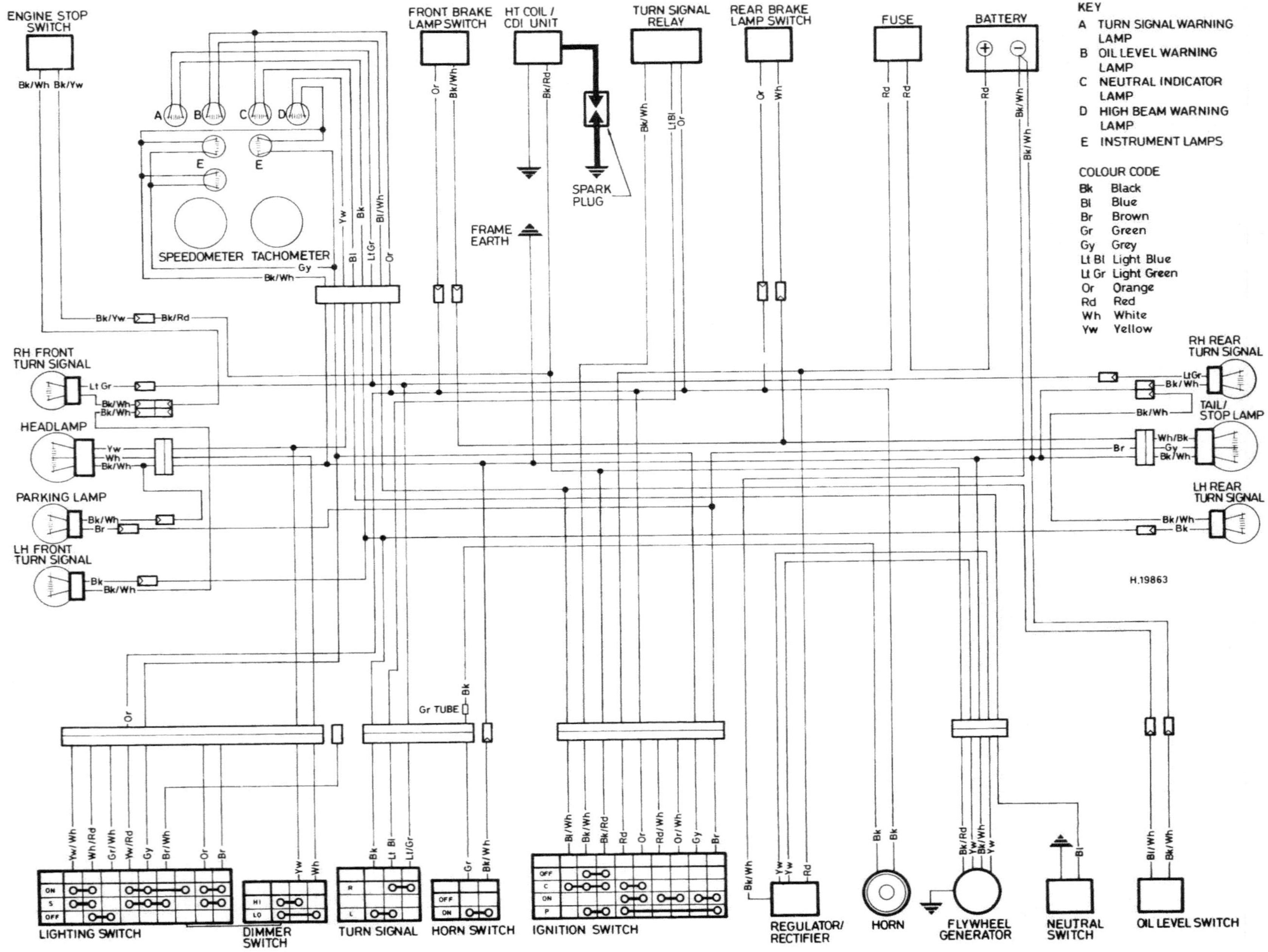

KEY

A TURN SIGNAL WARNING LAMP
B OIL LEVEL WARNING LAMP
C NEUTRAL INDICATOR LAMP
D HIGH BEAM WARNING LAMP
E INSTRUMENT LAMPS

COLOUR CODE

Bk Black
Bl Blue
Br Brown
Gr Green
Gy Grey
Lt Bl Light Blue
Lt Gr Light Green
Or Orange
Rd Red
Wh White
Yw Yellow

H.19863

RH REAR TURN SIGNAL
TAIL/ STOP LAMP
LH REAR TURN SIGNAL

OIL LEVEL SWITCH

NEUTRAL SWITCH

FLYWHEEL GENERATOR

HORN

REGULATOR/ RECTIFIER

IGNITION SWITCH

HORN SWITCH

TURN SIGNAL

DIMMER SWITCH

LIGHTING SWITCH

BATTERY

FUSE

REAR BRAKE LAMP SWITCH

TURN SIGNAL RELAY

HT COIL / CDI UNIT

SPARK PLUG

FRAME EARTH

FRONT BRAKE LAMP SWITCH

SPEEDOMETER   TACHOMETER

ENGINE STOP SWITCH

RH FRONT TURN SIGNAL

HEADLAMP

PARKING LAMP

LH FRONT TURN SIGNAL

Gr TUBE

Wiring diagram

Conversion factors

Length (distance)

Inches (in)	X	25.4	= Millimetres (mm)	X	0.0394	= Inches (in)
Feet (ft)	X	0.305	= Metres (m)	X	3.281	= Feet (ft)
Miles	X	1.609	= Kilometres (km)	X	0.621	= Miles

Volume (capacity)

Cubic inches (cu in; in³)	X	16.387	= Cubic centimetres (cc; cm³)	X	0.061	= Cubic inches (cu in; in³)
Imperial pints (Imp pt)	X	0.568	= Litres (I)	X	1.76	= Imperial pints (Imp pt)
Imperial quarts (Imp qt)	X	1.137	= Litres (I)	X	0.88	= Imperial quarts (Imp qt)
Imperial quarts (Imp qt)	X	1.201	= US quarts (US qt)	X	0.833	= Imperial quarts (Imp qt)
US quarts (US qt)	X	0.946	= Litres (I)	X	1.057	= US quarts (US qt)
Imperial gallons (Imp gal)	X	4.546	= Litres (I)	X	0.22	= Imperial gallons (Imp gal)
Imperial gallons (Imp gal)	X	1.201	= US gallons (US gal)	X	0.833	= Imperial gallons (Imp gal)
US gallons (US gal)	X	3.785	= Litres (I)	X	0.264	= US gallons (US gal)

Mass (weight)

Ounces (oz)	X	28.35	= Grams (g)	X	0.035	= Ounces (oz)
Pounds (lb)	X	0.454	= Kilograms (kg)	X	2.205	= Pounds (lb)

Force

Ounces-force (ozf; oz)	X	0.278	= Newtons (N)	X	3.6	= Ounces-force (ozf; oz)
Pounds-force (lbf; lb)	X	4.448	= Newtons (N)	X	0.225	= Pounds-force (lbf; lb)
Newtons (N)	X	0.1	= Kilograms-force (kgf; kg)	X	9.81	= Newtons (N)

Pressure

Pounds-force per square inch (psi; lbf/in²; lb/in²)	X	0.070	= Kilograms-force per square centimetre (kgf/cm²; kg/cm²)	X	14.223	= Pounds-force per square inch (psi; lbf/in²; lb/in²)
Pounds-force per square inch (psi; lbf/in²; lb/in²)	X	0.068	= Atmospheres (atm)	X	14.696	= Pounds-force per square inch (psi; lbf/in²; lb/in²)
Pounds-force per square inch (psi; lbf/in²; lb/in²)	X	0.069	= Bars	X	14.5	= Pounds-force per square inch (psi; lbf/in²; lb/in²)
Pounds-force per square inch (psi; lbf/in²; lb/in²)	X	6.895	= Kilopascals (kPa)	X	0.145	= Pounds-force per square inch (psi; lbf/in²; lb/in²)
Kilopascals (kPa)	X	0.01	= Kilograms-force per square centimetre (kgf/cm²; kg/cm²)	X	98.1	= Kilopascals (kPa)
Millibar (mbar)	X	100	= Pascals (Pa)	X	0.01	= Millibar (mbar)
Millibar (mbar)	X	0.0145	= Pounds-force per square inch (psi; lbf/in²; lb/in²)	X	68.947	= Millibar (mbar)
Millibar (mbar)	X	0.75	= Millimetres of mercury (mmHg)	X	1.333	= Millibar (mbar)
Millibar (mbar)	X	0.401	= Inches of water (inH₂O)	X	2.491	= Millibar (mbar)
Millimetres of mercury (mmHg)	X	0.535	= Inches of water (inH₂O)	X	1.868	= Millimetres of mercury (mmHg)
Inches of water (inH₂O)	X	0.036	= Pounds-force per square inch (psi; lbf/in²; lb/in²)	X	27.68	= Inches of water (inH₂O)

Torque (moment of force)

Pounds-force inches (lbf in; lb in)	X	1.152	= Kilograms-force centimetre (kgf cm; kg cm)	X	0.868	= Pounds-force inches (lbf in; lb in)
Pounds-force inches (lbf in; lb in)	X	0.113	= Newton metres (Nm)	X	8.85	= Pounds-force inches (lbf in; lb in)
Pounds-force inches (lbf in; lb in)	X	0.083	= Pounds-force feet (lbf ft; lb ft)	X	12	= Pounds-force inches (lbf in; lb in)
Pounds-force feet (lbf ft; lb ft)	X	0.138	= Kilograms-force metres (kgf m; kg m)	X	7.233	= Pounds-force feet (lbf ft; lb ft)
Pounds-force feet (lbf ft; lb ft)	X	1.356	= Newton metres (Nm)	X	0.738	= Pounds-force feet (lbf ft; lb ft)
Newton metres (Nm)	X	0.102	= Kilograms-force metres (kgf m; kg m)	X	9.804	= Newton metres (Nm)

Power

Horsepower (hp)	X	745.7	= Watts (W)	X	0.0013	= Horsepower (hp)

Velocity (speed)

Miles per hour (miles/hr; mph)	X	1.609	= Kilometres per hour (km/hr; kph)	X	0.621	= Miles per hour (miles/hr; mph)

Fuel consumption*

Miles per gallon, Imperial (mpg)	X	0.354	= Kilometres per litre (km/l)	X	2.825	= Miles per gallon, Imperial (mpg)
Miles per gallon, US (mpg)	X	0.425	= Kilometres per litre (km/l)	X	2.352	= Miles per gallon, US (mpg)

Temperature

Degrees Fahrenheit = (°C x 1.8) + 32   Degrees Celsius (Degrees Centigrade; °C) = (°F - 32) x 0.56

*It is common practice to convert from miles per gallon (mpg) to litres/100 kilometres (l/100km), where mpg (Imperial) x l/100 km = 282 and mpg (US) x l/100 km = 235

Index